SpringerBriefs in Business Process Management

Series editor

Jan vom Brocke, Vaduz, Liechtenstein

For further volumes:
http://www.springer.com/series/13170

Amy Van Looy

Business Process Maturity

A Comparative Study on a Sample of Business Process Maturity Models

Amy Van Looy
Faculty of Economics and Business
Administration
Ghent University
Ghent
Belgium

ISSN 2197-9618 ISSN 2197-9626 (electronic)
ISBN 978-3-319-04201-5 ISBN 978-3-319-04202-2 (eBook)
DOI 10.1007/978-3-319-04202-2
Springer Cham Heidelberg New York Dordrecht London

Library of Congress Control Number: 2013958134

Printed on acid-free paper

Springer is part of Springer Science+Business Media (www.springer.com)

To my family and friends
'On ne voit bien qu'avec le cœur. L'essentiel est invisible pour les yeux'
(Antoine de Saint-Exupéry, Le Petit Prince)

Acknowledgments

This book is an extract from my dissertation, which I submitted to the Faculty of Economics and Business Administration, Ghent University (Belgium), in fulfilment of the requirements for the degree of Doctor in Applied Economics. The work was successfully defended on 18 December 2012.

I would like to express my gratitude to the members of my doctoral committee for carefully reading my Ph.D. thesis and for the useful comments:

- Prof. Dr. Marc De Clercq (Dean-President, Ghent University)
- Prof. Dr. Patrick Van Kenhove (Academic Secretary, Ghent University)
- Prof. Dr. Geert Poels (Advisor, Ghent University)
- Prof. Dr. Manu De Backer (Advisor, University College Ghent, Ghent University, University of Antwerp)
- Prof. Dr. Monique Snoeck (KU Leuven)
- Prof. Dr. Frederik Gailly (Ghent University)
- Prof. Dr. Kevin McCormack (SKEMA Business School—Sophia Antipolis, The National Graduate School—Massachusetts)
- Prof. Dr. Ir. Hajo Reijers (Eindhoven University of Technology)

I truly appreciate their willingness to make the time available for the internal and public defence, and to trigger interesting discussions that provided me with constructive feedback. Thank you for generously investing time and energy in evaluating my work, giving me tips for publishing and sharing insightful ideas regarding my future research.

I particularly offer my sincerest gratitude to my advisory committee (Prof. Dr. Geert Poels, Prof. Dr. Manu De Backer, and Prof. Dr. Monique Snoeck) for their support along the way, for their trust in me and genuine interest in my research. I am deeply indebted to them for their help.

Finally, I am grateful to the University College Ghent for the financial support.

Contents

Summary

Business processes are at the heart of each organisation by describing its way of working. Globalisation, higher competitiveness, more demanding customers, growing IT possibilities, compliancy requirements, etc., all these challenges put pressure on organisations to perform better, and thus to obtain mature (or excellent) business processes. Consequently, business process maturity models (BPMMs) have been designed to help organisations gradually assess and improve their business processes. Some well-known examples are CMMI or OMG-BPMM, but many others exist.

In fact, so many BPMMs are around these days that organisations cannot see the wood for the trees. They risk selecting a BPMM that does not fit their needs, or a BPMM that might be of lower quality. Moreover, academics have frequently criticised maturity models for being consultancy speech, as those models typically simplify the complex reality. This criticism is primarily explained by the limited scientific knowledge on business process maturity, and how to measure and improve it. Particularly, we encountered three open knowledge issues: (1) scarce literature on BPMMs (indicating an unexplored research domain), (2) no comprehensive comparative studies on BPMMs (in contrast to the huge number of existing BPMMs) and (3) no theories of business process maturity (in contrast to the criticism). Nevertheless, the essential process improvements are not easy to realise and organisations may need some practical guidance on their journey towards excellence. Consequently, BPMMs remain important tools to help organisations, but they must be thoroughly examined to allow a critical view on the many BPMMs and to theoretically underpin them. For this purpose, the dissertation provides a comprehensive overview of existing BPMMs (N = 69). By including non-academic BPMMs, our sample is more comprehensive than other BPMM overviews. Further on, by including different process types (i.e. generic business processes, supply chains and collaboration processes), our sample suggests versatility which should facilitate transferability of our findings to other process types.

To properly investigate the problem statement, this dissertation covers the following three studies:

(1) a literature study to define the main concepts related to BPMMs, in order to obtain a common understanding to start from (issues 1 and 2);

(2) a classification study to identify and organise the possible capability areas and maturity types of BPMMs, in order to strengthen the BPMM foundation (issues 1, 2 and 3);
(3) a selection study to identify the criteria to be considered when choosing a BPMM out of the wide array, in order to derive evaluation scores for BPMMs and to provide practical advice on BPMM selection (issues 1 and 2).

The different studies constitute a comparative study based on a large sample of existing BPMMs (N = 69). To obtain theoretical insights into the concept of business process maturity, the studies are underpinned by theories and conceptual frameworks on maturity models, business processes and organisation management.

First, the literature study defines the scope, the terminology and the design of BPMMs. Maturity models deal with business processes if they deal with the following areas: (1) process modelling, (2) process deployment, (3) process optimisation, (4) process management, (5) a process-oriented culture and (6) a process-oriented structure. This BPMM scope is derived from clear and accepted definitions for a business process (BP), business process management (BPM) and business process orientation (BPO). Unique to this study is that we present a funnel structure in which the first two areas are attributed to BP maturity, the first four are attributed to BPM maturity and all six are attributed to BPO maturity. Translated to the BPMM terminology, the six areas are called business process capability areas, or areas of related skills and competences that are needed for a business process to perform excellently. Particularly, maturity levels indicate the overall growth throughout all areas addressed by a BPMM, whereas capability levels indicate the growth of an individual area. The higher the levels, the better business processes can be executed. As such, business process maturity expresses the expected business process performance based on the actual capabilities, in order to predict the actual performance of business processes. Finally, a BPMM is considered more usable if it provides both an assessment method, i.e. to assess the current levels, and an improvement method, i.e. to suggest improvement activities towards the desired levels. The findings are combined in a BPMM definition, which is used throughout this dissertation.

Secondly, the classification study elaborates on the six capability areas from the literature study by decomposing them into 17 sub areas. The sub areas are theoretically identified by drawing on established theories, and empirically validated by mapping them to our BPMM sample. Particularly, the sub areas for the first three areas are underpinned by theories of the traditional business process lifecycle, whereas the sub areas for the other three areas are more supported by organisation management theories, like performance and change management, human resource management and strategic management. Nevertheless, the BPMMs within our sample do not necessarily address all theoretically identified capability areas. More specifically, cluster analysis and discriminant analysis statistically reveal three maturity types being measured by our BPMM sample:

(1) BPM maturity, primarily focusing on business process modelling (1), deployment (2), optimisation (3) and management (4);

(2) BPO maturity, combining BPM maturity with a process-oriented culture (5) and structure (6);
(3) intermediate BPO maturity, limiting BPO maturity to some process-oriented aspects, usually cultural (5).

By adding the number of business processes to which BPMMs literally refer, it comes down to nine different maturity types: BPM maturity for one, more or all business processes, intermediate BPO maturity for one, more or all business processes, and BPO maturity for one, more or all business processes in the organisation. This extended BPMM classification allows critically evaluating the completeness of BPMMs, because BPO models are more complete than intermediate BPO models, which in turn are more complete than BPM models. Similarly, models for all business processes are more complete than models for specific business processes. The findings of this classification study are summarised into a conceptual framework of business process capability areas. As such, the study responds to the lacking consensus in the literature on capability areas that are necessary for reaching business process excellence. Moreover, evidence is given for diversity among the business process maturity types.

Thirdly, the requested critical view on BPMMs is continued in the selection study. It broadens our dissertation towards other BPMM characteristics than the capability coverage alone. This study results in 14 criteria that must be strategically thought through when selecting a BPMM.

(1) Which capability areas must be assessed and improved according to your needs?
(2) Must the BPMM define a roadmap per capability area, a roadmap for overall maturity, or both?
(3) How much guidance must the BPMM give on your journey towards higher maturity?
(4) Must the BPMM be generic (i.e. for business processes in general) or domain-specific (e.g. for business processes in supply chains or collaboration situations)?
(5) Which type of data must be collected during an assessment?
(6) How must information be collected during an assessment?
(7) For which purpose must the BPMM be used?
(8) Must evidence be explicitly given that the BPMM is able to assess maturity and helps to enhance the efficiency and effectiveness of business processes?
(9) How many business processes must be assessed and improved?
(10) How long must a particular assessment maximally take?
(11) Must the assessment questions and corresponding level calculation be publicly available (instead of only known to the assessors)?
(12) Must the BPMM explicitly recognise to include people from outside the assessed organisation as respondents in the assessment?
(13) How many questions must be maximally answered during an assessment?
(14) Must the BPMM be free to access and use?

These criteria are obtained from an international Delphi study with 11 BPM academics and 11 BPM practitioners, each from five continents. The findings are summarised into an online decision tool, called BPMM Smart-Selector, which is freely available at http://smart-selector.amyvanlooy.eu/. It concerns a questionnaire that helps making the necessary trade-offs when choosing amongst BPMMs. The tool is tested with three real-life case studies in both profit and non-profit organisations, with or without BPMM experience, and by both business and academic users. As the quality of our BPMM Smart-Selector also depends on the proposed BPMMs, two evaluation scores are calculated per BPMM: (1) a selection score based on the weights of the 14 final criteria, as assigned by the Delphi study and (2) a transparency score based on the completeness of BPMM design documents in order to be directly usable after selection. The transparency score copes with all criteria considered in the Delphi study (i.e. also those criteria that did not reach consensus to be excluded from the tool, neither to be included in the tool, and thus remain important to some extent). After this quality check, 60 out of the 69 collected BPMMs are made available in the tool.

Finally, avenues for future research are proposed for (1) continuing our theory-building approach on business process maturity, (2) enhancing existing BPMMs and (3) designing new BPMMs for cross-organisational business processes.

Chapter 1
Introduction

Abstract This chapter clarifies the context in which the research was conducted. By elaborating on the long history of business processes, it becomes an indispensable resource for management students, researchers, and practitioners. The essential facts of the industrial revolutions and quality management schools are combined with the official start of the process literature in the 1990s, characterised by reengineering and workflows. Also the influence of software engineering is discussed to explain the origin of maturity models, based on the tradition of CMMI, ISO/IEC 15504, FAA-iCMM and OMG-BPMM. Besides orienting readers to the book, the given background allows identifying a problem statement with knowledge issues in the current process literature. Although business process maturity models (BPMMs) turn out to be important aids for organisations, they are frequently criticised because many models exist without a theoretical foundation. This leads to the main research question for this book, requiring a methodological approach that follows behavioural-science and design-science paradigms. Throughout the studies covered in this book, theoretical findings will be altered with a content analysis of an unique sample of existing BPMMs. Particularly, this chapter presents a large overview of 69 BPMMs for generic processes, supply chains and collaboration processes, including references to be used by the readers.

Keywords Business process · Quality management · Business excellence · Business process reengineering · Process improvement · Capability · Maturity · Maturity model · Behavioural-science paradigm · Design-science paradigm · Content analysis

The overall purpose of this book is to gain new scientific insight into business process maturity. Before going into detail, we present three statements to clarify the practical relevance of our research topic:

- business processes are important;
- mature (i.e. excellent) business processes are important;
- business process maturity models (BPMMs) are important.

A. Van Looy, *Business Process Maturity*, SpringerBriefs in Business Process Management, DOI: 10.1007/978-3-319-04202-2_1,

These statements are primarily meant to introduce the reader to the topic of business process maturity. They are also backed up by an historical overview of business processes and business process maturity models (BPMMs), which should facilitate linking the context of this book to the reader's own experience. After getting acquainted with the research topic, we will demonstrate that the related practical problems give rise to significant knowledge problems, resulting in academic research needs and opportunities to create knowledge.

Statement 1: business processes are important

> '*Improving business processes*' was a top 10 business expectation of CIOs since its introduction to the CIO Agenda survey in 2005 until 2011. In 2012, it is combined with analytics/business intelligence '*to create new capabilities*' (N = 2,335) (Gartner 2012).

> '*Improving effectiveness and cost efficiency of business processes*' are two of the top three IT priorities of executives to improve business performance since the 2008 Global Survey (N = 927) (McKinsey 2011).

Business processes reflect the way of working in organisations, or how organisations operate (Harrington 2006). For instance, they explain how organisations produce products or deliver services, how the supply chain functions, or how departments or organisations collaborate. An example of a business process is visualised in Fig. 1.1.

Figure 1.1 shows which steps are needed to operate, in which sequence and by whom they are executed. Nonetheless, a business process is more than the order in which things are done, and thus differs from a workflow. Particularly, a workflow is rather the deployable version of a business process with operational details or '*the automation of a business process, in whole or part, during which documents, information or tasks are passed from one participant to another for action, according to a set of procedural rules*' (WfMC 1999, p. 8). On the contrary, a business process takes a concept view (not a realisation view) by also describing the high-level business context with the business process purpose and corresponding performance targets, etc. (Kannengiesser 2008; Weske 2010; zur Muehlen and Ho 2006). The scope of this book is thus broader than (technical) workflows.

The importance of business processes is already widely accepted in both information technology (IT) environments and business environments. First, successful information systems (IS) are assumed to support business processes and provide business-specific functionality. Hence, business processes must be documented at an early stage of a software development project. They are positioned at the beginning of frameworks regarding software engineering and the enterprise architecture, such as the Rational Unified Process or the Zachman framework (McGovern et al. 2004; vom Brocke and Rosemann 2010). Research on business processes can, for instance, investigate the relationship between business and technical artefacts, such as between process models and process-aware information systems. The present work, however, mainly focuses on the business environment by taking a management perspective. Particularly, business processes claim a central position in quality standards, business excellence models, and performance

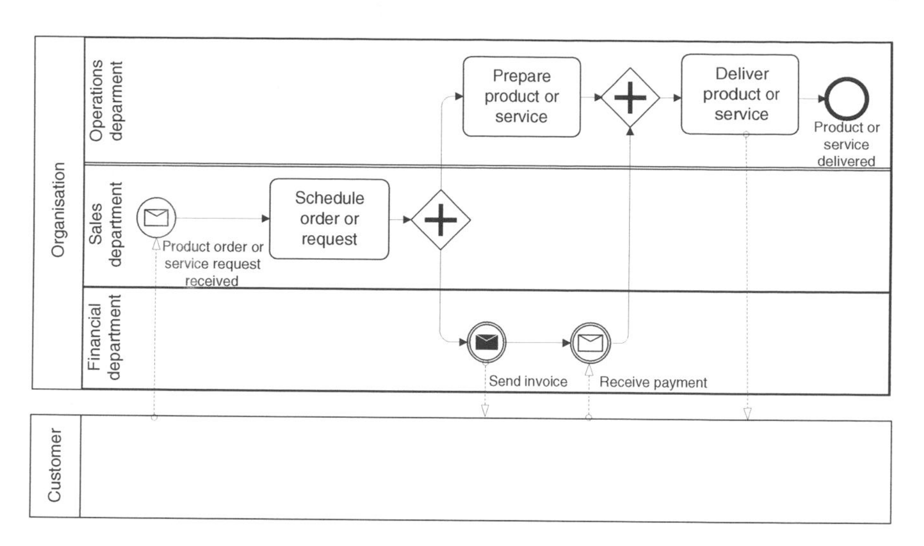

Fig. 1.1 A business process example, modelled in the BPMN process language

improvement techniques, such as the ISO 9000-series (ISO 2009), the European Foundation for Quality Management (EFQM) (EFQM 2010), or the Balanced Scorecard (Kaplan and Norton 2001). Quality means meeting (technical and customer) specifications, whereas performance also involves effectively and efficiently meeting strategic objectives (van Dooren et al. 2004). Quality standards (ISO 2009) and business excellence models (EFQM 2010) stimulate private and public organisations to document, follow and regularly audit their business processes. Further on, the balanced scorecard of Kaplan and Norton (2001) measures organisational performance, with business processes being one of the four measurement perspectives. Another management approach is described by the value chain theory of Porter (1985). A value chain is an end-to-end business process that describes an organisation's activities in a particular industry, i.e. a business unit instead of an individual department. It consists of dependent processes and their interactions through an organisation, i.e. from suppliers to customers. The involved business processes contain primary activities, e.g. core processes in manufacturing or marketing, or secondary activities, e.g. support processes in human resource management. A value chain focuses on creating value, i.e. the amount that customers are willing to pay for a product or service, while performing cheaper or better than its competitors. It is defined as '*a general framework for thinking strategically about the activities involved in any business and assessing their relative cost and role in differentiation*' (Porter 1985, p. xv).

Furthermore, the importance of business processes is reflected in the existence of specialised international journals (e.g. the Business Process Management Journal since 1995) and conferences (e.g. the Business Process Management Conference since 2003). Moreover, business process management (BPM) courses are institutionalised at several universities, also at the University College Ghent and Ghent University, and training institutions, like Cevora or SAI in Belgium. According to Houy et al. (2010), this evidence proves that managing business processes is '*not a temporary fashion, but an evolving trend in management science*' (p. 620).

Statement 2: mature (i.e. excellent) business processes are important

> *Good is no longer good enough. Doing the right things "right" is not good enough. Having the highest quality and being the most productive doesn't suffice today. To survive in today's competitive environment, you must excel* (Harrington 2006, p. xix).

Nowadays, organisations are increasingly focusing on their business processes, i.e. on their way of doing business, in order to excel. This means that they strive for the highest level of performance. For instance, Harrington (2006) presents BPM as one of his five pillars to improve an organisation's performance towards business excellence (i.e. besides project, change, knowledge, and resource management). The link between business processes, quality and performance is, however, not a novelty. Only the focus has now changed from performance to business (process) excellence, i.e. the highest level of performance. This pursuit of excellence is mainly due to: (1) higher customer expectations in the globalised market, and (2) growing IT possibilities to support business processes (Harrington 2006; vom Brocke and Rosemann 2010). Particularly, today's customers have higher requirements than ever before. In a growing globalised market, private organisations are striving to excel in order to gain competitive advantage. Similarly, customers have translated their high expectations towards public organisations, and prompt them to outperform in their societal obligations. Furthermore, as the possibilities of IT continue to grow, business processes are enabled to cross the borders of departments and even organisations. Also the trend of e-business, e-commerce and e-government stimulates process automation and integration, and thus collaboration between departments and organisations.

BPM can contribute to quality and business excellence by assuring an uniform way of working and by continuously looking for optimisations (Beckford 2009; Cobb 2003). Merely modelling and deploying a business process does not imply that it is also an excellent one, or at least a good one. A business process should also be continuously optimised and managed. This brings us to the notion of maturity, which is a measure to indicate how well business processes can perform (Harrington 2006; Maier et al. 2009). Maturity aims at systematically improving the capabilities of business processes and their organisations to deliver higher performance over time. It relates to the expected performance, which is an indicator of the actual performance (Hammer 2007; McCormack 2007c; Rosemann and de Bruin 2005a). To our knowledge, the current literature acknowledges two types of business process maturity (de Bruin and Rosemann 2007; Pöppelbuss and Röglinger 2011): (1) the maturity for the management of specific business processes, based on CMMI, e.g. the OMG example in Fig. 1.2, and (2) the maturity of BPM, i.e. the management of all business processes in an organisation.

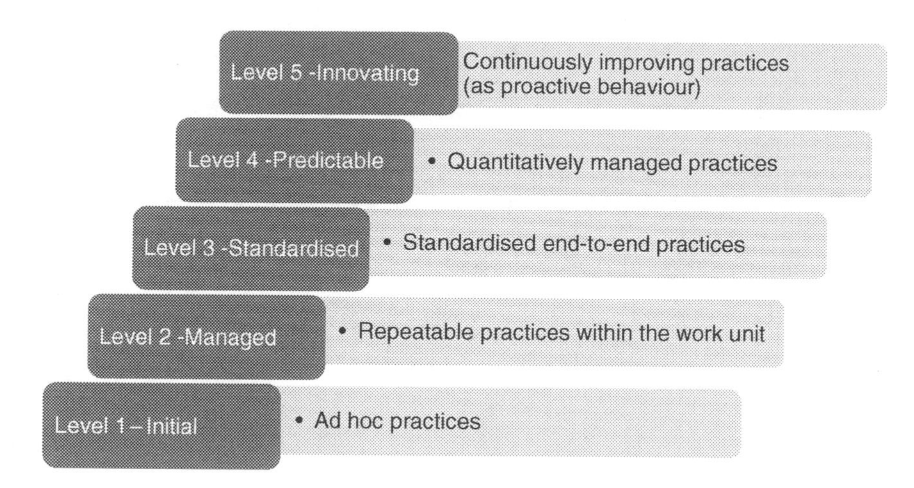

Fig. 1.2 A BPMM example (OMG 2008)

Statement 3: business process maturity models (BPMMs) are important

> *Too few organizations understand how good their processes can be because they have no measure of how bad they are* (Harrington 2006, p. 99).

> *A maturity model assumes that progress comes in stages, ultimately reaching an end goal* (McCormack 2007c, p. 73). *Defining the end goal of this journey and finding out where you are on this journey is critical* (McCormack 2007a, p. 61).

Although mature business processes are important, process improvements are not easy to realise. Therefore, various authors have proposed a business process maturity model (BPMM) to gradually improve business processes towards higher maturity. They present a sequence of lifecycle levels and a step-by-step roadmap with goals and best practices to reach each consecutive level. BPMMs can be compared with a navigation system or GPS, which explain step-by-step best practices to reach a goal, i.e. to reach business (process) excellence. Your itinerary or step-by-step roadmap consists of a sequence of planned stops, i.e. lifecycle levels, while explaining the criteria to reach each consecutive stop, i.e. goals and best practices. Hence, BPMMs are practical tools to assist organisations on their journey towards business (process) excellence. An example is given in Fig. 1.2.

In general, a maturity model (MM) is a tool to systematically assess and improve capabilities or critical success factors to reach a goal. Translated to BPMMs, it concerns the capabilities of business processes and their organisations to reach business (process) excellence. Capabilities are competencies (e.g. skills and knowledge) to achieve the targeted results, i.e. the ability to perform, or the expected performance of a business process. Examples of possible capabilities for business processes are:

- the capability to accurately model a business process as it is currently performed;
- the capability to monitor the actual process execution;
- the capability to identify deviations between the actual process execution and the prescribed process execution;
- the capability to set key performance indicators (KPIs) for achieving process goals;
- the capability to measure process goal achievement; etc.

Related capabilities are collected into capability areas. Two types of lifecycle levels exist: maturity levels and capability levels. Maturity levels indicate the growth through all capability areas together. Sometimes, capability levels are present to indicate the growth for each capability area separately, instead of an overall growth.

Conform to the two known types of maturity (de Bruin and Rosemann 2007; Pöppelbuss and Röglinger 2011), an organisation with ten business processes can choose between: (1) a BPMM that assesses and improves each process separately (maturity type 1), i.e. separate maturity and capability levels for each of the ten processes, (2) a BPMM that assesses and improves the organisation's mastery in BPM (maturity type 2), i.e. one maturity level and one capability level per capability area for all ten processes, or (3) a BPMM that combines both alternatives (maturity type 1 and 2). This distinction may, for instance, indicate that the organisation is generally capable of modelling its processes, but that only some core processes already have a graphical design, whereas other use textual descriptions, or that support processes do not need the same maturity level as core processes.

The ongoing assessment and improvement of business process maturity is considered to be fundamental to the realisation of business (process) excellence. For instance, Rosemann (2010) recommends the use of a BPMM to determine an organisation's BPM strategy with '*at its core a roadmap that specifies the planned BPM-related activities over the next 3–5 years*' (p. 268). Furthermore, higher maturity levels facilitate networking with other organisations (McCormack 2007a; Pesic 2009). Nevertheless, BPMMs do not pretend being the only right way of handling. For instance, not all business processes must necessarily reach the highest maturity level, e.g. business processes that do not directly contribute to the core business. Additionally, the desired maturity level may depend on contextual factors, like the degree of market competitiveness. Plattfaut et al. (2011) explain that '*the achievement of highly mature BPM requires enormous resources specifically dedicated to process change. These resources cause costs for both initial set-up and maintenance. However, when the market environment is not dynamic and, thus, does not demand quick responses to environmental changes, these costs might not pay off*' (p. 9). Consequently, organisations must determine their desired level by taking into account the organisational objectives and context. As a result, the maturity levels and capability levels should rather be optimised than maximised, i.e. the most appropriate level to achieve the organisational objectives (de Bruin and Rosemann 2007; Harrington 1991; Tolsma and de Wit 2009). From this point of view, the presence of capability levels becomes more interesting. Capability areas can be separately improved, risking suboptimal results on the one hand, but also allowing organisations to choose which areas are less relevant to their situation.

The remainder of this introductory chapter is structured as follows. We first summarise the history of business processes (Sect. 1.1) and BPMMs (Sect. 1.2). Section 1.3 elaborates on the problem statement with needs and opportunities for new scientific knowledge, followed by the research question and objectives in Sect. 1.4. Next, the research methodology (Sect. 1.5) and the data collection phase (Sect. 1.6) are explained. Afterwards, we present the research outline of this book (Sect. 1.7).

1.1 History of Business Processes (= Quality Management)

This book contributes to a rather new research domain. The modern business process literature is claimed to have started with the value chain theory of Porter (1985) and business process reengineering (BPR) in the late 1980s and the early 1990s (Hammer 1996; Houy et al. 2010; vom Brocke and Sinnl 2011). In the next section, we will argue that it is accompanied by a tradition of software process improvements, e.g. (Curtis et al. 1992; Kellner et al. 1999). As from the 1990s, business processes are also frequently studied as (technical) workflows, e.g. (Jablonski and Bussler 1996). Moreover, vom Brocke and Sinnl (2011) argue that BPR's initial technology focus is only recently broadened towards management-related topics. The early roots, however, can be traced back to the industrial revolutions and the upcoming factories. Business processes were originally investigated by organisation management domains, particularly from quality management and performance management perspectives. For instance, business process modelling and optimisation are quality management techniques to obtain products and services that effectively and efficiently meet customer requirements (Beckford 2009). BPMMs serve as a complement to quality standards (ISO 2009) or business excellence models (EFQM 2010) by elaborating on business process improvements. They gain importance as business processes have become paramount throughout the decades, as shown in Fig. 1.3.

Hammer and Champy (2003) and Chang (2006) refer to Adam Smith and his prototypical pin factory as the start of quality management and business process thinking. In his work 'Wealth of Nations' (Smith and Mazlish 2002), originally published in 1776, Smith described the different steps (i.e. simple tasks or operations) to produce a pin. He also argued that more pins could be produced when performed by multiple workers, each specialised in a subset of tasks. This work is situated in the first industrial revolution, which was characterised by labour specialisation and fragmentation of work.

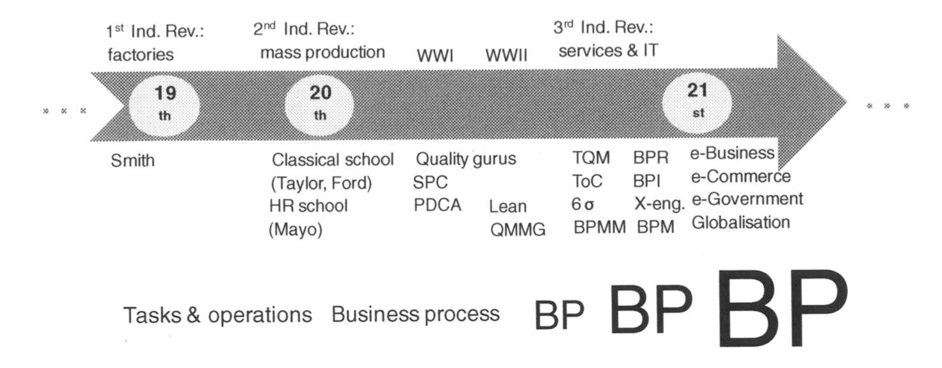

Fig. 1.3 The increasing importance of business processes throughout the centuries

Nevertheless, the roots of quality management and business process thinking are more frequently traced back to the second industrial revolution, i.e. with mass production and assembly lines in the early twentieth century (Beckford 2009; Ross and Omachonu 2004; Fox and Frakes 1997). This period was characterised by two management schools: (1) the classical management school (e.g. Taylor, Ford), which saw the organisation as a machine and separately analysed the functionality of each part, and (2) the human relations management school (e.g. Mayo, Maslow), which took an organic view on the organisation and analysed the interactions between the individual parts. Especially Taylor's scientific management with time-and-motion studies of manufacturing tasks are worth mentioning as a predecessor of business process thinking (Ross and Omachonu 2004).

The (soft) human relations management school reacted to the (hard) classical management school by emphasising the importance of human capital. Despite their opposing views, both early management schools still focused on internal improvements. Beckford (2009) explains that current quality thinkers, and thus also business process thinkers, are still inspired by both schools, albeit supplemented now by customer and environmental needs. Meanwhile, quality management has evolved towards organisation management, namely (Thijs and Bouckaert 2007):

- from producer orientation towards customer orientation;
- from defect detection towards defect prevention:
  - inspection: verifying the conformance of outputs to technical requirements;
  - quality control: inspection combined with statistical methods to measure quality at the end of the production line;
  - quality assurance: preventing defects by introducing a quality program, in which organisational processes are modelled and with clear responsibilities;
  - total quality management: continuously improving end-to-end processes, products and services to achieve the organisational goals more efficiently and effectively;

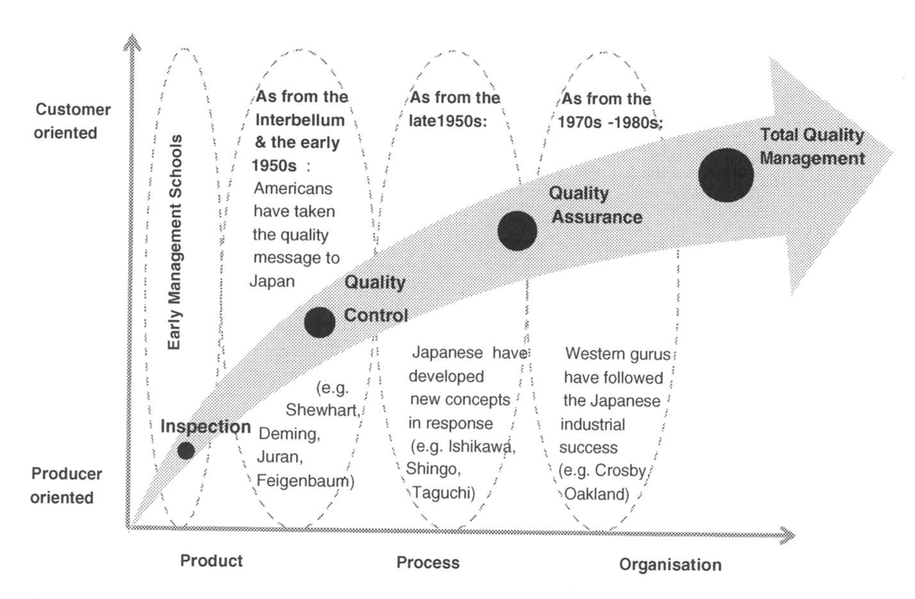

Fig. 1.4 The evolution in classical quality thinking (Thijs and Bouckaert 2007; Beckford 2009)

- from responsibility of the management and the quality department towards responsibility of the whole organisation;
- from simple controlling techniques towards advanced quality management techniques and models.

This evolution was spread throughout the twentieth century, as shown in Fig. 1.4.

Classical quality thinkers were inspired by the successful post-war economic revival in Japan (Beckford 2009; Chang 2006). Although a discussion of the quality gurus is out-of-scope, we draw attention to:

- Shewhart (1980, 1986): the first quality guru, albeit on manufacturing processes, and founder of the statistical process control (SPC) for Six Sigma (6σ);
- Deming (1994, 2000): who broadened Shewhart's ideas to business processes within any department, and founder of the Plan-Do-Check-Act (PDCA) process lifecycle;
- Crosby (1979): who created the quality management maturity grid (QMMG), i.e. the first maturity model to our knowledge.

As from the late 1980s and the early 1990s, the importance of business processes significantly increased. Hammer (1996) explains that a shift was made towards end-to-end processes, i.e. crossing different departments. From that moment, authors are referring to modern business process literature (vom Brocke and Sinnl 2011). Particularly, four contemporary quality movements emphasise

Table 1.1 The four contemporary quality evolutions in business process thinking

How to improve BPs?	BPR (early 1990s: Hammer and Champy)	BPI (mid-1990s: Harrington)	X-engineering (2002: Champy)	BPM—third wave (2002: Smith and Fingar)
Radically	X		X	
Incrementally		X		X
Cross-departmentally	X	X		
Cross-organisationally			X	X
Use of IT	X	X	XX	XX

end-to-end business processes: (1) BPR, (2) business process improvement (BPI), (3) business process X-engineering, and (4) the third wave of BPM. As summarised in Table 1.1, the first and third movement are based on radical process improvements, whereas the second and fourth movement focus on incremental improvements. It thus concerns two pairs of action and reaction, i.e. one in the 1990s and the other in the 2000s. Given the evolutions in society, the first two movements concentrate on end-to-end business processes within a single organisation in the 1990s, whereas the last two movements emphasise cross-organisational integration in the 2000s. All movements make use of the growing IT possibilities since the third industrial revolution, but with a strong mandatory character for the movements of the 2000s, i.e. X-engineering and the so-called third wave of BPM.

In the late 1980s and the early 1990s, Western companies were facing higher costs, higher customer demands and more competition with foreign products. Therefore, the first generation of BPR aimed at decreasing costs, so that jobs should not be exported to low-wage countries. IT must be used to redesign business processes, i.e. starting from a clean slate, instead of merely automating existing processes (Hammer 1990; Hammer and Stanton 1995). Besides IT, a variety of existing quality management techniques was used, e.g. flowcharting, statistical process control, benchmarking, etc. (Attaran 2004; Davenport 1993; Coombs and Hull 2001). Although BPR was promising, it received a lot of criticism. Studies have shown that most BPR projects failed, e.g. due to merely downsizing or outsourcing (Beckford 2009), introducing incompatible IT systems across departments (Basu and Palvia 2000; O'Neill and Sohal 1999), generic best practices without differentiating the organisation (Chang 2006), managers not giving up control of their functional areas (Attaran 2004), etc.

During the mid-1990s, a new movement gained importance as a reaction to BPR: BPI which advocated incremental improvements for end-to-end business processes (Harrington 1991; Harrington and Harrington 1995; Harrington et al. 1997). This movement conforms to the mainstream in quality thinking and to three currently wide-spread process improvement programs (Nave 2002): theory of constraints (ToC), Lean, and Six Sigma (6σ). Although BPR quickly lost its credibility (Carr 2003; Chang 2006), BPR and BPI still co-exist.

The second generation of BPR started as X-engineering in the 2000s. Champy (2002) recognises that his early BPR view with cost reductions was in the advantage of shareholders, not customers or employees: *'fewer people were asked to do more work, while others laid off and customer service deteriorated'* (p. 2). Therefore, his second BPR generation involves all stakeholders. *'X-engineering is the art and science of using technology-enabled processes to connect businesses with other businesses and companies with their customers to achieve dramatic improvements in efficiency and create value for everyone involved'* (Champy 2002, p. 3). He argues that reengineering remains necessary due to the ongoing international competitiveness and work redundancy, but adapted to the new business climate, i.e. the digital age. This evolution in business process thinking was stimulated by the emerging Internet use of organisations, which resulted in a trend towards e-business, e-commerce, and e-government in the late 1990s and early 2000s. Managers must now cross "X" organisational boundaries, and reinvent processes across partnering organisations to face the new challenges of connectedness and interdependency in an electronic business-to-business (B2B) environment. IT, especially the Internet, is seen as the most important enabler for business process improvements (Champy 2002; Attaran and Attaran 2004).

As in the 1990s, an incremental counterpart emerged, like BPI, but which was also in favour of cross-organisational integration by using IT, like X-engineering. It is presented by Smith and Fingar (2002, 2006) as the third wave of BPM. It focuses on automated and agile processes in the new information society, which is characterised by globalisation and customisation. Their third wave follows Taylor's scientific management of the 1920s (first wave—BPM as an analysis of non-automated methods and procedures) and the reengineering tradition of the 1990s (second wave—BPM as a radical, one-time endeavour). It uses the same variety of techniques as BPI (e.g. Lean, Six Sigma, brainstorming, cause-effect diagrams, flowcharting, etc.), albeit with an additional focus on management and automation (Delphi Group 2002; Chang 2006). Open standards and universal process modelling languages (e.g. BPMN) are presented as the integration solution for business applications. Furthermore, a BPM suite or system (as an example of a process-aware information system) is proposed by the authors, i.e. a technical engine of software tools for directly modelling, deploying and optimising business processes by business people instead of software engineers (Smith and Fingar 2002, 2006; Windle 2004).

Nowadays, the business process literature aims at increasing organisational performance by (technology-enabled) process improvements (Burlton 2001). It searches for a balance between radical and incremental changes, depending on the context (Chang 2006; Zhao and Cheng 2005). For instance, once a process is reengineered, continuous improvements will be necessary (Attaran 2004). The current literature thus profits from both opposites by taking a more refined approach. Although modelling and optimising business processes are still frequently addressed as technical or hard issues (e.g. like workflows), the soft aspects or the human side of business change should not be underestimated. For instance, business processes must be permanently managed by professional process owners, whether successive improvements are radical or incremental (Burlton 2001). Furthermore, various authors assert that process improvements must also be supported by the underlying organisational characteristics (like the organisational culture and structure with, among others, policies and standards) in order to realise higher organisational performance (McCormack and Johnson 2001; de Bruin and Rosemann 2007; vom Brocke and Sinnl 2011). It thus requires a broader business process orientation (BPO) instead of being limited to BPM as such.

1.2 History of BPMMs (= Software Engineering)

The previous section explained the growing need for business process improvements. We now turn to the history of BPMMs, i.e. models that provide direct guidance to realise those essential process improvements. BPMMs have a shorter tradition than business processes, starting with software engineering in the 1980s. Nevertheless, BPMMs are also strongly influenced by quality thinking. For instance, statistical process control is used on the higher maturity levels (Shewhart 1986), and the PDCA process lifecycle refers to gradually reaching those higher maturity levels (Deming 2000). Furthermore, the early maturity model of Crosby (1979) is the precursor of BPMMs, albeit for quality management.

The need for BPMMs increased due to a software crisis that started in the 1970–1980s. Many IT projects faced great difficulties in delivering on time, on budget and within scope. Consequently, the first maturity models for software development processes emerged (Humphrey 1989; SEI 1987). Such processes are indeed the core business processes for software organisations. For instance, a maturity model can be used to assess the software development capabilities of a (potential) contractor, i.e. to evaluate the risks of contracting, and requiring a minimum maturity level before doing business. Inspired by these maturity models for software development processes, the first maturity models for generic business processes appeared in the late 1990s and the early 2000s, i.e. either by designing new models or by integrating existing models for specific process types (Ahern et al. 2004). To our knowledge, there are four main tracks within the BPMM literature (Table 1.2): (1) CMMI, (2) ISO/IEC 15504, (3) FAA-iCMM, and (4) OMG-BPMM. They are generic BPMMs, developed in parallel, and widely accepted and used in various sectors. They are also supported by large communities, i.e. government or government-sponsored institutions, international standards bodies or consortia with mixed membership (Ahern et al. 2004; de Bruin and Rosemann 2007; Lee et al. 2009; Paulk 2008; Rohloff 2009b).

In the 1980s, Crosby's maturity model was adopted by IBM to introduce quality practices into software development projects. As from 1986, Humphrey (an IBM representative) assisted the Software Engineering Institute (SEI) in developing a series of maturity models, in collaboration with other organisations from industry

Table 1.2 The main tracks in the BPMM history

	1980s	1990s	2000s
SEI[a]	• Process maturity framework (1987)	• CMM (1991, 1993) • CMM variants	• CMMI (2000, 2002, 2006, 2010)
ISO/IEC		• SPICE (1995) • ISO/IEC TR 15504 (1998)	• ISO/IEC 15504 (2003, 2008)
FAA		• iCMM (1997, 2001, 2004)	
OMG			• BPMM (2008)

[a] CMM is a registered trademark, and CMMI a service mark of the Carnegie Mellon University

and government. SEI is a research centre at the Carnegie Mellon University, funded by the US Department of Defence (DoD). Their first initiative was the Process Maturity Framework, and used a maturity questionnaire with questions related to specific maturity levels (Humphrey 1989; SEI 1987, 1993). This framework is the predecessor of the Capability Maturity Model for Software (CMM or SW-CMM), which was released in 1991 and 1993. Instead of a questionnaire, SW-CMM defined key process areas (KPAs) with goals and best practices per maturity level to facilitate establishing a process improvement program (SEI 1993; Paulk et al. 1995). The KPAs are examples of capability areas in this book. After 1993, a large variety of specialised CMMs was derived from the structure of SW-CMM, e.g. for systems engineering (SE-CMM), software acquisitions (SA-CMM), systems security engineering or human resources (People CMM). Sheard (1997) summarised the multitude of process improvement models and standards for software development, given that '*developers may need to consider a daunting number of frameworks. The field truly is a quagmire in which process improvement efforts can bog down if an organisation is not careful*' (p. 2). In 2000, the Capability Maturity Model Integrated (CMMI) was launched to integrate all CMM variations and to facilitate extensions to other domains (Ahern et al. 2004; Kulpa and Johnson 2008; Curtis et al. 2001). The first version of CMMI was still limited to software engineering and systems engineering (SEI 2000a, b). A major turning point was reached in 2002, when CMMI also introduced processes related to integrated product and process development, and supplier sourcing outside the software context (SEI 2002a, b). As from version 1.2, CMMI's structure has changed into best practices for the development of software or other products (SEI 2006b), acquisitions (SEI 2007) and services in general (SEI 2009). The scope of CMMI thus gradually evolved from the software sector to the economic sector in general. To our knowledge, CMMI is the best known maturity model that is applicable to different types of business processes, and can now be genuinely called a BPMM for generic business processes. It still inspires the creation of many other BPMMs, which follow the overall increased need for process improvements in order to excel. Nevertheless, a comprehensive overview of models addressing generic business processes (similar to Sheard's 1997 framework quagmire for software development processes) is still lacking.

The second main track is developed by the International Organisation for Standardisation (ISO). In 1993, the SPICE project (Software Process Improvement and Capability dEtermination) was launched to build an international standard for assessing software development processes, in collaboration with SEI (Paulk et al. 1995). In 1995, the SPICE drafts were accepted as a technical report: ISO/IEC TR 15504 (ISOSPICE 2010; SQI 2007; Simon et al. 1997). In 2003, it became a full international standard. Instead of being constrained to software development processes, ISO/IEC 15504 now supports different process types for also non-software organisations (ISO/IEC 2003; 2004a, b, c; 2008).

The third main track originates from the US Federal Aviation Administration. Based on the CMM variants, FAA released its integrated capability maturity model for engineering, management and acquisition processes in 1997. In 2001, its scope broadened to all CMMI processes (FAA 2001). Nowadays, FAA-iCMM is used by organisations that rely on complex IT systems for service delivery (Ibrahim and Pyster 2004). In contrast to the other main tracks, FAA-iCMM supports additional application areas that translate the specific practices to certain disciplines, such as safety and security systems (FAA 2004).

Finally, the Object Management Group (OMG) released a generic BPMM in 2008 (OMG 2008). OMG is a consortium that creates open standards for IT, among others process modelling. OMG argues that its BPMM focuses on cross-departmental business processes for any organisation type, whereas the other main tracks focus more on specific domains or projects related to software development and maintenance projects (Curtis and Alden 2007; Alden 2007).

Figure 1.5 visualises that all main tracks in the BPMM literature are organised in a similar way, i.e. with main business process capability areas decomposed in sub areas, each with specific and generic goals and practices.

Figure 1.5 shows that, although the terminology may slightly differ, all main tracks in the BPMM literature have a series of goals and best practices within the so-called key process areas or KPAs (i.e. equivalents of capability areas in our research). They evaluate maturity in accordance to the sequence and the number of goals achieved (Hammer 2007). Goals and practices are (1) specific or unique to each business process type, or (2) generic for all processes to get institutionalised or ingrained into the way of working. For instance, specific goals for acquisition processes relate to supplier selection, contracts, invoices, etc. On the other hand, these processes get institutionalised by generic practices, like modelling, planning, training, performance measurement, etc. Most tracks provide both specific and generic goals and practices. Only ISO/IEC 15504 relies on external process frameworks to define the specific goals and practices, i.e. by reusing other main tracks, international standards (e.g. ISO/IEC 12207 for software lifecycle processes) or non-ISO process frameworks (e.g. Automotive SPICE, Banking SPICE, Medi SPICE, SPICE for SPACE (aerospace), Enterprise SPICE (cross-departmental processes), etc.) (ISOSPICE 2010; Rout et al. 2007). Best practices propose what could be done to satisfy the goals. Consequently, organisations can still decide to use other methods and approaches, e.g. other best practices models like COBIT, ITIL, PMBOK, Prince2 and the ISO 9000-series.

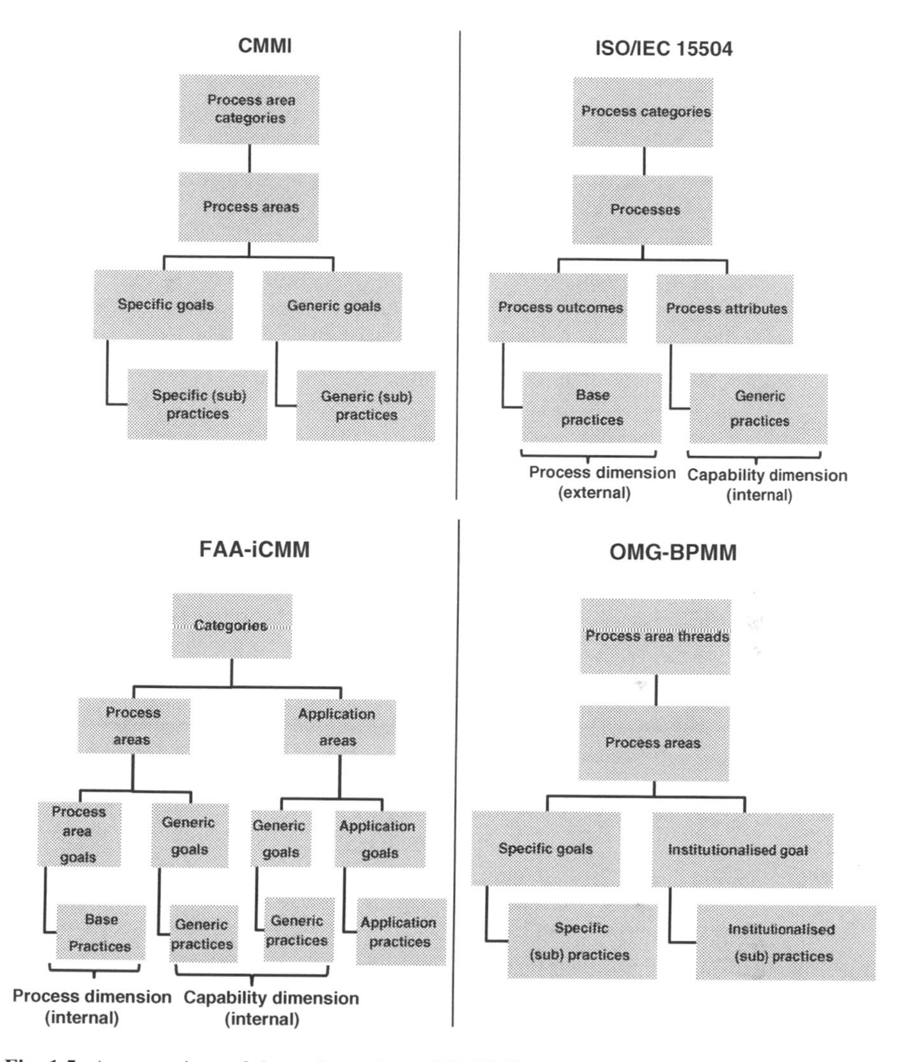

Fig. 1.5 A comparison of the main tracks in BPMM history

Like the evolution in process thinking, BPMMs have broadened their initial scope towards all kinds of business processes for all kinds of organisations. For instance, human or soft aspects are reflected since People CMM, and various integrated BPMMs or BPMMs for generic processes have been developed. Besides focusing on particular business processes, more and more BPMMs start focusing on the BPM mastery of an organisation, i.e. all its business processes (Rosemann and de Bruin 2005b). Furthermore, a variety of maturity models do not focus on organisational processes, but rather look at organisational domains, practices or assets, like project management, e-government or the IT architecture (Buglione 2009). The scope of the research presented in this book is limited to maturity models which explicitly focus on the business processes of organisations.

1.3 Problem Statement

Besides the practical relevance of BPMMs, the historical background allows us to become aware of practical implications. As process improvements are essential, the main tracks co-exist with a wide range of other (mostly non-academic) BPMMs. This myriad makes it difficult to determine the most appropriate BPMM in function of organisational characteristics and objectives. Organisations risk selecting a BPMM that does not fit their needs, or selecting a BPMM that might be of lower quality. In our search for solutions, however, we encountered three open knowledge issues regarding BPMMs: (1) scarce literature, indicating an unexplored research domain, (2) no comprehensive comparative studies, and (3) no business process maturity theories. Genuinely addressing these issues is of paramount importance for both the academic world and practitioners wishing to adopt a BPMM.

Issue 1: scarce literature on BPMMs
To illustrate the importance of the business process literature, we searched for related articles in the Web of Science Index (i.e. containing multiple citation databases) and the Business Process Management Journal. The BPM Journal is not ranked as an A1 journal at Ghent University, and therefore not included in the citation databases. Nevertheless, it is a recognised journal within the BPM research domain. An initial search was conducted for articles until 2009, i.e. when starting the work. Mid-2012, the same search was repeated to verify any changes when finalising the work.

We first discuss the initial search for articles until 2009 to properly reflect the gap that we faced at the beginning of our research. Table 1.3 shows a total of 8,525 articles when searching for "business process" in the full text, whereas 1,601 articles for "business process management" when starting the research. In this book, we will differentiate BPM from "business process orientation" (BPO). We will argue that BPM refers to characteristics of business processes, i.e. related to the traditional business process lifecycle, whereas BPO adds organisation-specific characteristics to BPM, i.e. to make the organisational culture and structure more process-oriented. The notion of BPO was proclaimed by McCormack and Johnson (2001), among others, but rather as a synonym for BPM. In our study, we will explicitly distinguish BPM from BPO in order to emphasise a difference in scope. As this refinement is not yet adopted in the business process literature, only 34 articles were found for "business process orientation" in the full text until 2009.

Unlike the high number of articles on business process (management), significantly less academic writing existed on business process maturity until 2009. Table 1.3 shows 155 articles when searching for "business process" combined with "maturity" in the full text. Many of these articles, however, merely mention "maturity" once or twice throughout the text or use it in another context, e.g. to indicate that a research domain has reached maturity. Therefore, we further refined our results by searching for "business process" in the full text and "maturity" in the title. This resulted in 19 articles, of which many discuss maturity models that

Table 1.3 The number of academic articles related to the research topic (until 2009 and mid-2012)

Number of articles until 2009 (start) until mid-2012 (end)	SCI-EXP[a]	SSCI[b]	A&HCI[c]	CPCI-S[d]	CPCI-SSH[e]	BPM journal	Total
"business process" in text	1,873	775	4	4,359	965	549	8,525
	2,330	1,027	6	5,108	1,085	672	10,228
	(+457)	**(+252)**	(+2)	**(+749)**	**(+120)**	**(+123)**	**(+1,703)**
"business process management" in text	235	42	0	641	134	549	1,601
	344	90	0	819	147	672	2,072
	(+109)	**(+48)**	(=)	**(+178)**	**(+13)**	**(+123)**	**(+471)**
"business process orientation" in text	7	5	0	8	3	11	34
	9	11	0	11	3	24	58
	(+2)	**(+6)**	(=)	**(+3)**	(=)	**(+13)**	**(+24)**
• "business process" in text and	19	13	0	30	7	86	155
	29	25	0	40	8	130	232
• "maturity" in text	**(+10)**	**(+12)**	(=)	**(+10)**	**(+1)**	**(+44)**	**(+77)**
• "business process" in text and	3	4	0	7	3	2	19
	7	11	0	9	4	4	35
• "maturity" in title	**(+4)**	**(+7)**	(=)	**(+2)**	**(+1)**	**(+2)**	**(+16)**
• "business process" in title and	1	2	0	5	2	1	11
	2	5	0	7	2	2	18
• "maturity" in title	**(+1)**	**(+3)**	(=)	**(+2)**	(=)	**(+1)**	**(+7)**

[a] Science citation index—expanded; [b] Social sciences citation index; [c] Arts & humanities citation index; [d] Conference proceedings citation index—science; [e] Conference proceedings citation index—social science & humanities

do not primarily focus on business processes, but rather focus on business-IT alignment, interoperability, knowledge management, project management, etc. Finally, in order to obtain only articles on business process maturity, we searched for "business process" and "maturity" in the title. Only 11 articles met these criteria by focusing on a particular BPMM or on the relationship of a particular BPMM with performance. The limited BPMM literature can be explained by being a rather new academic research domain. BPMMs are mostly studied within the context of specific process types (e.g. software development processes) or by industry (e.g. white papers). Becker et al. (2010) assert that the academic interest in maturity models is, however, increasing.

We repeated the search at the end of July 2012 to verify how much the business process literature has increased during the last years. Table 1.3 shows that most new articles deal with business processes in general or business process management, respectively with 1,703 and 471 additional articles. 24 new articles are added for business process orientation, which means an increase of almost 70 % compared to 2009. This increase indicates that scholars become more acquainted with the BPO concept. Although the BPMM literature remains scarce compared to the larger business process literature, the statement of Becker et al. (2010) is confirmed. Seven new articles refer to "business process" and "maturity" in the title, which means an increase of almost 60 % compared to 2009. Nevertheless, it also reconfirms the research opportunities regarding the present research topic.

Issue 2: no comprehensive comparative studies on BPMMs

Given the importance of mature (i.e. excellent) business processes and in spite of the scarcity of academic literature on the topic, a proliferation of BPMMs exists (Paulk 2004; Sheard 2001). Some estimates exceed 200 process improvement frameworks, including BPMMs and standards (Curtis and Alden 2007; El Emam and Birk 2000; Moore 1999). Another estimate, restricted to BPMMs, mentions over 150 models addressing one or more areas of BPM (de Bruin and Rosemann 2007).

Nevertheless, the many existing BPMMs are assumed to differ in quality (Rosemann 2010). To our knowledge, a comprehensive BPMM overview or comparison including all their design elements was lacking at the start of this work. Limited attempts at such studies have been made by Lee et al. (2007), and Rosemann and de Bruin (2005a). More extended overviews are presented by Hüffner (2004), and Maier et al. (2012). Hüffner (2004) compared 11 maturity models related to TQM or BPM, before creating a maturity model. Maier et al. (2012) examined 24 maturity models, albeit not limited to business processes and not elaborating on the capability areas. None of these authors primarily intended to offer a comprehensive study on a large number of academic and non-academic BPMMs, in order to obtain a state-of-the-art. As a result, practitioners have no overview of existing BPMMs or the substantial differences, which makes an informed choice difficult when starting the process excellence journey. Besides organisations, also academics who want to apply a single BPMM in their research risk not seeing the wood for the trees.

Simultaneously with this work, Pöppelbuss and Röglinger (2011) and Röglinger et al. (2012) prepared a small overview of 10 BPMMs (which are also considered in this book). Nevertheless, their research is situated in the design science literature. Instead of being limited to design principles for maturity models, our research primarily aims to gain insight into business process maturity by using the whole plethora of BPMMs and to provide selection advice on how to choose the right BPMM. A detailed comparison between the authors' design criteria and our selection advice is provided within a journal article resulting from this work. It is argued that BPMM selection requires more than just design considerations (for instance, also practical or financial considerations).

Issue 3: no theories of business process maturity
To our knowledge, a theory of business process maturity is lacking, which makes the BPMM research open to criticism. For instance, maturity models are frequently criticised as consultancy speech by academics because they oversimplify complex issues (Maier et al. 2009; Plattfaut et al. 2011; Röglinger et al. 2012). Indeed, BPMMs inherently balance the complex reality with practical guidance. Nonetheless, they suggest a systematic sequence of improvements, frequently based on best practices of other organisations, instead of improving ad hoc. Therefore, Maier et al. (2009) argue that maturity models are '*widely used within industry as an evaluative and comparative basis for improvement*' (p. 8). Moreover, some scholars even predict that the overall adoption of maturity models is likely to increase (Scott 2007).

The criticism is mainly tackled in three ways. First, some scholars examine the effect of business process maturity on business (process) performance. Their empirical research suggests that projects or organisations with a higher maturity level appear to perform better (Jiang et al. 2004; Harter and Slaughter 2000; Lockamy III and McCormack 2007; McCormack et al. 2008; Skrinjar et al. 2008; Skrinjar et al. 2007). Such empirical research is usually conducted to validate a particular BPMM. It has the potential to build a theory if the effect can be repeatedly shown independent of the use of a particular BPMM.

Secondly, other scholars try to tackle the criticism by taking a contingency approach, i.e. by adapting a maturity model to organisation-specific characteristics. For instance, maturity models that suggest different improvements based on the organisation sector (e.g. public or private), size (e.g. large-sized or SMEs), type (e.g. for products of services), or environment (e.g. dynamic or stable). This issue is also called 'situationality', and is still under research (Hain 2010; Plattfaut et al. 2011; Plomp and Batenburg 2010; Ravesteyn and Jansen 2009; van Steenbergen et al. 2010). It may result in the creation of new BPMMs, even though many BPMMs already exist. The question remains if such refinements are strong enough to overcome the criticism without a strong theoretical and empirical foundation.

Thirdly, maturity models have been mainly studied from a design science perspective, focusing on the process of designing a maturity model. Many scholars propose specific phases to properly create a maturity model. They generally cover pre-design, design and post-design (sub) phases (Becker et al. 2009;

de Bruin et al. 2005; Maier et al. 2009; Mettler and Rohner 2009; Tapia et al. 2008; van Steenbergen et al. 2010). Additionally, the seven criteria of Hevner et al. (2004) for evaluating the design of information systems (IS) artefacts have been translated towards maturity model design by Becker et al. (2009). These phases with evaluation criteria may be seen as a theory of the design of maturity models (or rather as a design process or methodology). When designed accordingly, BPMMs are supposed to have a sound methodological foundation. Nevertheless, the design theory remains general for all maturity models. It does not translate specific design elements to BPMMs, e.g. the capability areas to assess and improve lifecycle levels.

We are convinced that a theory of business process maturity is required, as a fourth way to overcome the criticism. Building and testing a theory has the potential to lift BPMMs out of the non-academic atmosphere and launch it as a full research domain. For instance, a theory could explain what factors related to the concept of business process maturity contribute to business (process) performance. Hence, a theory could support the design of BPMMs that foster the enhancement of these factors. Such new knowledge would add refinements to previous studies that already showed a significant effect on performance by using a particular BPMM. Furthermore, it could enlighten different dimensions of maturity being measured by the BPMMs, and how they relate to each other. The general need for theory building within the business process literature is also confirmed by Houy et al. (2010). Given the limited scientific knowledge on business process maturity, the present work addresses this third knowledge issue as preparatory work. Particularly, we will elaborate on business process maturity as expected performance, while empirically investigating the link with actual performance is part of future research. Our contribution to this issue includes gathering preliminary information that helps define the problem domain and suggest hypotheses for further research. We will not build and test a theory ourselves, but we will examine how business process maturity is underpinned by existing theories. This BPMM foundation remains poorly explained so far, as most BPMMs rely only on empirical studies. As such, the present work can be seen as a first, essential step towards a theory of business process maturity.

1.4 Research Question and Objectives

Our research question is motivated by the three open knowledge issues, i.e. (1) scarce literature indicating a yet unexplored research domain, (2) no comprehensive comparative studies in contrast to the huge number of existing BPMMs, and (3) no theories of business process maturity in contrast to the criticism.

RQ. How can business process maturity models (BPMMs) be defined, classified, and selected?

The research is exploratory in nature, as the concept of business process maturity was not yet clearly defined at the beginning of this work. The main objective is to gain significant scientific insight into business process maturity by reviewing the available literature, reviewing existing BPMMs, and discussions with BPM experts. It results in a comparative study on a sample of existing BPMMs that explains how we can define, classify and select BPMMs.

Objective for issue 1: contributing to the BPMM literature
The first issue, i.e. scarce BPMM literature, will be implicitly addressed by reporting on our research findings at international conferences and in academic journals. Further on, we conclude the book by presenting future research avenues. Moreover, with our research, we also hope to stimulate other researchers to delve into business process maturity.

Objective for issue 2: defining, classifying, and selecting BPMMs
We will primarily compare and analyse BPMMs in order to create a BPMM classification and to provide selection advice. Due to the large number of existing BPMMs, our intention is not to design yet another BPMM for generic business processes. Instead, we fully acknowledge prior BPMM efforts by conducting a large-scale comparative study on existing BPMMs, i.e. issue 2. After defining the main concepts in our research, we will start by comparing the business process capability areas, which are core, and add other design characteristics afterwards. By first comparing the capability areas, we tend to create a BPMM classification allowing a meaningful distinction among the many BPMMs. With the comparison of additional characteristics, we tend to provide selection advice by means of a BPMM decision tool that helps users choose a BPMM that best fits their needs.

Objective for issue 3: grounding the BPMM literature
The BPMM classification mentioned in the previous objective will be underpinned by theory to strengthen the BPMM foundations, before investigating the effect on actual performance in future work. Hence, besides summarising and structuring for issue 2, we also want to ground the BPMM literature to some degree, i.e. issue 3. The BPMM classification will be based on a conceptual framework of business process capability areas. This framework can contribute to a theoretical framework on business process maturity in future research. Therefore, empirical findings will be carefully elicited, following a sound methodology, and explained by existing (process or organisational) theories, if possible.

1.5 Research Methodology

The methodology used in the different studies of this book will be described in Chap. 2. Nonetheless, we hereby present a general overview of our research approach. BPM is an interdisciplinary research field across applied computer sciences (e.g. technology engineering and information systems) and applied social sciences (e.g. economics and business administration) (vom Brocke and Rosemann 2010; Weske 2010). Various authors assert that two research paradigms dominate the domains of applied sciences (Simon 1996), such as BPM (Houy et al. 2010) or information systems (IS) (Hevner et al. 2004; March and Smith 1995; Walls et al. 2004; Winter 2008), namely:

- the behavioural-science paradigm (i.e. building and testing theories to explain and predict situations as existing phenomena in the natural world);
- the design-science paradigm (i.e. building and testing artificial objects and phenomena, called new artefacts, to solve problem situations, or human intervention in the natural world).

Both paradigms are paralleled by 'meta' research, respectively social sciences and design science in general to guide researchers and ensure rigour. To differentiate the applied design science from the 'meta' design science, Winter (2008) respectively refers to 'design research' and 'design science' (both contributing to 'design science research'). The paradigms are also complementary, as insight into problem situations facilitates artefact development to solve those situations, and vice versa (Hevner et al. 2004; Houy et al. 2010).

The BPMM literature is mainly restricted to the design-science paradigm, in which maturity models are analysed as artefacts. The IS design research cycle, guidelines and artefact types have been used in the context of maturity models. Table 1.4 visualises that scholars have translated articles on the IS design-science paradigm to the design-science paradigm of maturity models (and not specifically BPMMs). It also shows that developing particular BPMMs (as artefacts) belongs to the design-science paradigm of maturity models, albeit as design research instead of design science.

For instance, Table 1.4 explains that the IS design research cycle has been adapted to the process of designing maturity models. Furthermore, Hevner et al. (2004) formulate seven guidelines to evaluate the design of IS artefacts, which have also been translated towards the design of maturity models by Becker et al. (2009). Finally, Table 1.4 refers to the different IS artefact types of March and Smith (1995), which have been discussed for maturity models by Mettler and Rohner (2009) and van Steenbergen et al. (2010). These are examples of design science, and primarily focus on the process of developing maturity models in general (instead of BPMMs in particular). On the other hand, concrete BPMMs are artefacts designed by design research according to the design science principles for maturity models.

BPMM research in the behavioural-science paradigm remains less advanced. BPMMs implicitly rely on the effect of business process capability areas and maturity on business (process) excellence, as shown in Fig. 1.6. Current findings are, however, limited to studies on particular BPMMs which conduct behavioural research on the performance outcomes, e.g. (McCormack and Johnson 2001; Skrinjar et al. 2008).

To examine prior work, we started from a large sample of existing BPMMs (N = 69). As this sample is used throughout the research, Sect. 1.6 elaborates on the data collection phase. From the 69 collected BPMMs, 50 models take the

Table 1.4 The design-science paradigm in the BPMM literature

IS design-science paradigm	References	Translated towards: MM design-science paradigm	References
IS design research cycle	(Hevner et al. 2004; Peffers et al. 2007)	Pre-design, design and post-design (sub) phases = Design science	(Becker et al. 2009; de Bruin et al. 2005; Maier et al. 2009; Mettler and Rohner 2009; Tapia et al. 2008; van Steenbergen et al. 2010)
IS design research guidelines	(Hevner et al. 2004)	MM requirements = Design science	(Becker et al. 2009)
IS artefact types	(March and Smith 1995)	• MM as IS artefact types = Design science • Concrete BPMMs as artefacts = Design research	• (Mettler and Rohner 2009; van Steenbergen et al. 2010) • Sample (N = 69)

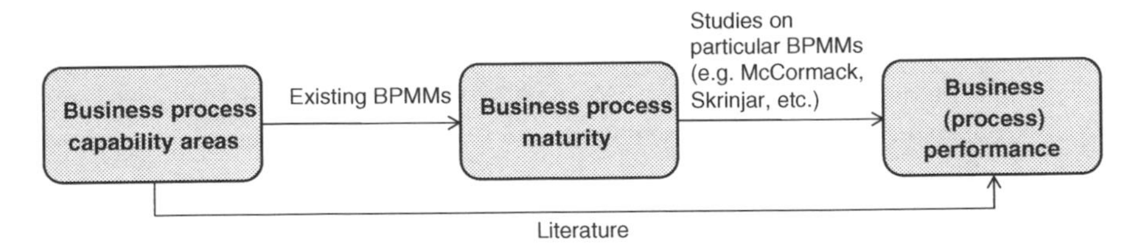

Fig. 1.6 The behavioural-science paradigm in the BPMM literature

performance outcomes for granted. Only 19 BPMMs empirically investigated the effect on performance after their use, e.g. by surveys, case studies, interviews, or trials. Moreover, only half of them also provide statistical evidence useful for generalisation. Such studies are, however, biased by the single BPMM used. To our knowledge, a study that simultaneously addresses different BPMMs is still lacking. For hypothesis testing, the BPMM foundation with underlying theories must be explored first, i.e. the direct effect of capability areas on performance.

Translated to our research, this book is situated in both research paradigms, namely:

- the behavioural-science paradigm for building and testing a BPMM classification to explain the capability areas that contribute to business process maturity (or expected performance);
- the design-science paradigm for building and testing a BPMM decision tool (as a new artefact) to choose the right BPMM.

We will compare and classify capability areas for theory building on business process maturity, particularly by contributing to the BPMM foundation. Afterwards, the comparative study is broadened towards all BPMM aspects to give advice on how to select a proper BPMM, i.e. by building and testing an online BPMM decision tool for practitioners. Consequently, we consider gaining insight into business process maturity as a first essential step to tackle criticism and to explore avenues for future research. To conduct this research, we adopted the IS research framework of Hevner et al. (2004) '*for understanding, executing, and evaluating IS research combining behavioural-science and design-science paradigms*' (p. 79). Our research cycle, as shown in Fig. 1.7, thus applies to both paradigms by (1) investigating a problem, (2) building a solution, e.g. a BPMM classification or artefact, based on existing knowledge and by specifying and validating candidate solutions before choosing one, (3) testing the solution with real-life data, and (4) evaluating its implementation, resulting in new knowledge. Interplay between theory and practice is supported to assure relevance to the business environment, and rigour by relying on academic knowledge throughout the cycle.

We subsequently discuss the research cycle for the planned BPMM classification and the BPMM decision tool. The research cycle is not applied to the first study, as the intended definitions are rather a prerequisite for classifying and selecting BPMMs.

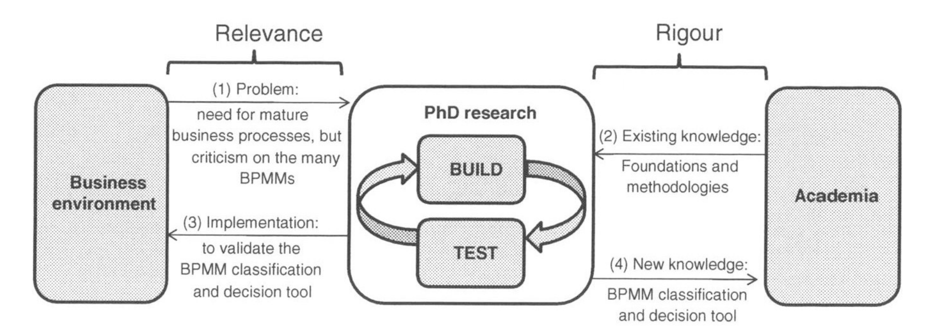

Fig. 1.7 The research cycle used in this book, based on Hevner et al. (2004)

- The BPMM classification will address all issues of the problem statement by enhancing the BPMM foundation (1). The underlying capability areas will be derived from the business process literature and established theories (2), and empirically tested by the BPMM sample (3). Afterwards, the BPMM classification will be built by exploratory classification techniques (2), and tested by confirmatory classification techniques (3). The resulting new knowledge concerns a conceptual framework of capability areas for business process maturity (4).
- The BPMM decision tool will address issues 1 and 2 of our problem statement by advising how to choose a BPMM out of the many BPMMs (1). Input for the tool will be derived from the BPMM sample and BPM experts based on high scholarly research standards (2), resulting in a prototype to be tested by case studies (3). New knowledge will be acquired on the trade-offs for BPMM selection, allowing to evaluate BPMMs accordingly, and on the methodology to create a BPMM decision tool (4).

By iteratively following the building and testing activities of our research cycle, Table 1.5 explains how we can contribute to three of the five theory types from Gregor's framework (2006).

A theory for analysing and describing BPMMs relates to our full comparative study by pointing out similarities and differences between BPMMs regarding their capability coverage (for the BPMM classification) and regarding other BPMM aspects (for selection advice in the BPMM decision tool). Secondly, a theory for explaining BPMMs is mainly covered in the classification study by understanding the capability areas that contribute to business process maturity (as critical success factors for performance). Finally, our selection advice and BPMM decision tool contribute to a theory for design and action by elaborating on how BPMMs must be chosen (as guidelines on how to properly select a BPMM). Finally, to address the other theory types of Gregor (2006), follow-up research will investigate to which degree business process maturity predicts actual performance outcomes, and which combinations of capability areas are more predictive than others.

Table 1.5 The theory types of Gregor (2006), translated to business process maturity

Types of theory	Causal explanations	Testable propositions (= hypotheses to predict)	Prescriptive statements (= guidelines to act)
(1) Theory for analysing and describing: "WHAT is?" e.g. a framework to classify BPMMs, which serves as an organising and simplifying device by pointing out similarities and differences between BPMMs	–	–	–
(2) Theory for explaining: "WHY?" e.g. understanding the capability areas that contribute to business process maturity	X	–	–
(3) Theory for predicting: "WHAT will be?" e.g. the power of business process maturity to predict business (process) performance (out-of-scope)	–	X	–
(4) Theory for explaining and predicting = (2) + (3) (out-of-scope)	X	X	–
(5) Theory for design and action: "HOW to?" e.g. guidelines to design a BPMM (out-of-scope) e.g. guidelines to select a BPMM	–	–	X

1.6 BPMM Sample (N = 69)

Throughout the research, a unique and large sample of existing BPMMs was used as input. This section explains how the BPMMs were sampled and analysed.

1.6.1 Sampling

Existing BPMMs were collected during the second quarter of 2010. We initially searched for articles in academic databases (i.e. SCI-Expanded, SSCI, A&HCI, CPCI-S, CPCI-SSH, BPM Journal) and search engines (i.e. Google, Google Scholar) by using the combined keywords of '*process*' and '*maturity*'. Then, we traced the references in the identified articles to get access to other relevant sources. BPMMs were included on two conditions: (1) the recognition of lifecycle levels for maturity and/or capability, which are typical for a maturity model, and (2) a primary focus on generic business processes (BP-focused), supply chains (SC-focused) or collaboration processes (collaboration-focused).

The first sampling condition ensured that only maturity models were included, and not just process standards, e.g. ITIL (OGC 2007), project standards, e.g. Prince2 (OGC 2010), or collaboration standards, e.g. RosettaNet and GS1 (CGF 2010; VICS 2004). Benchmarking studies were only selected if the results are directly linked to levels with a road map (AberdeenGroup 2006; Poirier et al. 2010; APQC 2007). Further on, we excluded process improvement techniques and diagnostic tools (Foggin et al. 2004), or frameworks defining different types of process transformations (Edwards et al. 2000), integration types (Frohlich and Westbrook 2001) or collaboration types (Holweg et al. 2005; Spekman and Carraway 2006).

The second sampling condition further narrowed the scope of this book towards BPMMs. Given the proliferation of BPMMs (Sheard 2001) and for reasons of generalisation, we initially searched for maturity models regarding generic business processes. We excluded BPMMs addressing specific process types, such as maturity models limited to concurrent engineering processes (Khalfan et al. 2001; Santanen et al. 2006), product design processes (Moultrie et al. 2007), IT management processes, such as COBIT (IT Governance Institute 2007) or Nolan (1973) and Renken (2004), and project management processes (Kwak and Ibbs 2002). However, models that integrate various domain-specific BPMMs were included in our sample to represent the specific process types to some degree, e.g. by the main tracks in BPMM literature. Also maturity models regarding supply chains and collaboration processes were selected to examine cross-organisational value chains. In a cross-organisational setting, value chains are also called value streams (Burlton 2010) or value systems (Weske 2010). A supply chain is a cross-organisational value chain for products and services among (strategically relevant) suppliers and customers, i.e. from the initial order to the ultimate consumption.

It mainly includes manufacturers, distributors, wholesalers, retailers and warehouses (Böhme 2008; Tan 2001). The supply chain focus is on collaborative planning, forecasting and replenishment, i.e. synchronising future activities (or scheduling the supply and forecasting the demand) and current activities (or inventory management and order management) (VICS 2004). Five supply chain processes are typically identified, also known as the standardised SCOR (or scorecard) model: plan, source, make, deliver and return (Supply-Chain Council 2008). On the other hand, a collaboration process coordinates the interaction between business entities, i.e. departments or individual organisations, as a repeatable set of exchange activities regarding messages and/or other material input and output (Bauer et al. 2006). Collaboration can be interpreted as: (1) data collaboration which refers to the interface between processes that links output to input, e.g. sending and receiving an order, or (2) task collaboration which implies a business process with shared activities, e.g. having a meeting (Goethals 2006). By including supply chains and collaboration processes, the sample acknowledges the evolution in process thinking, which is now also characterised by process integration and collaboration among external partners.

Additionally, we must note that maturity models were only included in our sample if they directly address business processes as a primary focus. A maturity model should not only mention business processes in its design documents. It should also aim at improving business processes throughout the traditional business process lifecycle, i.e. to support '*the design, administration, configuration, enactment, and analysis of business processes*' (Weske 2010, p. 5). Examples of excluded maturity models aim at improving business rules (Debevoise 2007) or strategic business-IT alignment (Luftman 2004), and thus only indirectly the business processes themselves. For instance, Luftman (2004) addresses cultural aspects to facilitate collaboration between process actors and the IT department, but without a link to the business process lifecycle. On the contrary, maturity models for operational business-IT alignment, like (Tapia et al. 2008), use IT to improve business processes, and were included.

We acknowledge some limitations regarding the accessibility of documents, the language (English or Dutch), and the keywords. Nonetheless, the technique turned out to be fruitful in terms of the number of BPMMs identified, which supports generalisation. A sample of 69 BPMMs was collected: (1) 37 BPMMs for generic business processes (13 academic and 24 non-academic), (2) 24 BPMMs for supply chains (9 academic and 15 non-academic), and (3) 8 BPMMs for process collaboration (6 academic and 2 non-academic). Academic BPMMs are developed and owned by scholars or academic institutions, whereas non-academic BPMMs are owned outside academia, for instance by consultancy or professional institutions. By including non-academic BPMMs, our sample is much larger than other overviews on BPMMs (Rosemann and vom Brocke 2010). Furthermore, by including different process types (i.e. generic, supply chains, collaboration), our sample suggests versatility which should facilitate transferability of our findings to other process types, e.g. software development processes. The latter is, however, not examined within the scope of this book.

Table 1.6 The most referenced BPMMs

Frequency	(%)	BPMM	Background	Mentioned by
31	44.9	SEI[a]	BP	AOU, BIS, BPM, CAM1, DEL, ESI2, FAA, FRA, GAR1, HAM, HAR1, HAR2, IDS, ISO, LEE, LMI, MAG, MCC2, NET, O&I, OMG, RAM, RIV, ROH, ROS, SAP, SKR, SMI, SPA, TAP, WOG
13	18.8	SCC	SC	BOH, CGR, CSC, EKN, HAR2, MCC2, MCL, NET, PMG, ROS, SIM, SPA, TOK
9	13.0	VIC	Collaboration	AND, BOH, CGF, CSC, EKN, MAN, NET, PMG, SCH2
6	8.7	ROS	BP	LEE, MAG, O&I, ROH, SPA, WIL
4	5.8	CSC	SC	CGR, MCC2, MCL, SIM
4	5.8	FIS	BP	LEE, MAG, ROH, ROS
4	5.8	HAM	BP	CAM2, O&I, ROH, SAP
4	5.8	HAR2	BP	LEE, MAG, ROS, SMI
4	5.8	ISO[a]	BP	ESI1, FAA, LEE, SEI
3	4.5	MCC2	SC	CAM3, NET, TAP (+ 2 references for the predecessor MCC1: SKR, WIL)
3	4.5	OMG[a]	BP	ESI2, LEE, ROH
3	4.5	PMG	SC	CAM3, CGR, MAN
3	4.5	SMI	BP	LEE, ROH, ROS
1	1.5	FAA[a]	BP	SEI

[a] Main track in the BPMM literature

The list of sampled BPMMs is available in appendix. For reasons of conciseness, the study refers to particular models through unique IDs. Each ID consists of the first three letters of the first author's name or abbreviation. If the letter combination already exists, then a number is added to obtain uniqueness.

The representativeness of our BPMM sample is illustrated in Table 1.6. It shows BPMMs to which other BPMMs mostly refer, and have thus gained credibility in this sense. To avoid arbitrariness, only BPMMs with three or more references are listed, as well as the exemplar models of the main tracks in the BPMM literature (Sect. 1.2).

Eight of the thirteen most referenced BPMMs are BP-focused, of which three are main track models: SEI, ISO and OMG. The fourth main track model, FAA, is only referred to once, but this by SEI (or the model to which most BPMMs refer). We recall from the BPMM history (Sect. 1.2) that these four main tracks in the BPMM literature are supported by large communities (Harmon 2009; Sheard 2001; Van Loon 2004). Except for FAA, Table 1.6 adds that they are also considered worthwhile by other BPMMs, i.e. important enough to mention. The same reasoning applies to SCC for supply chains and VIC for collaboration processes, i.e. also supported by large communities (respectively the Supply Chain Council and the Voluntary Interindustry Commerce Solutions Association) and being the second and third most referenced BPMM in our sample (Harmon 2010;

Skjoett-Larsen et al. 2003). The representativeness of our sample is further confirmed by containing other prominent BPMMs, such as ROS and HAM (Harmon 2009).

Finally, we must note that this book presents a comprehensive BPMM sample for research purposes. We do, however, not pretend that no other BPMMs exist in addition to the 69 collected BPMMs. We recall that sampling was done once (i.e. in the second quarter of 2010). Other BPMMs are likely to exist, which were not included in the sample due to the mentioned sampling limitations. Meanwhile, also new BPMMs might have been designed. However, at first sight, the new academic articles on business process maturity in Table 1.3 rather apply existing BPMMs than designing new ones. Nevertheless, an exact number of known BPMMs always remain a rough estimate at a certain moment in time.

1.6.2 Content Analysis

The design documents of the collected BPMMs were repeatedly analysed over time, beginning in the third quarter of 2010 until the second quarter of 2012 (without adding new BPMMs meanwhile). As such, our content analysis was refined as the research findings progressed, for instance when decomposing main business process capability areas into sub areas or when eliciting additional criteria for BPMM selection. At a later stage, new BPMMs can still be used to further validate the findings of this book (although, at first sight, we did not discover any new BPMMs while finalising this work).

We started from literature on maturity model design to interpret the design elements of a BPMM (Becker et al. 2009; de Bruin et al. 2005; Maier et al. 2009; Mettler and Rohner 2009; Tapia et al. 2008; van Steenbergen et al. 2010). These articles, as shown in Table 1.4, deal with maturity models in general and do not acknowledge the specific BPMM context. Some authors, like Pöppelbuss and Röglinger (2011) and Röglinger et al. (2012), also illustrate these general design elements by using a small sample of BPMMs, albeit without much detail. A Meta theory of BPMM design appeared to be lacking, as it turned out that elaborations were still required regarding the concrete options for each of the general design elements, e.g. to detail the specific lifecycle levels, the capability areas, etc. Consequently, the variables within each design element were identified by following similar coding stages as the 'Grounded Theory' (Glaser and Strauss 1967):

- initial (open) coding: we read the collected texts by constantly going back and forth to compare existing BPMM designs. Hence, we identified possible attributes and variables;
- intermediate (axial) coding: the attributes and variables were rethought and linked to the general design elements in the literature. It resulted in final variables for our research;
- advanced (selective) coding: we reread the collected texts to encode what is literally written in these texts to the obtained variables.

A qualitative research approach, i.e. content analysis, was chosen to get introduced to the main BPMM aspects (not necessarily limited to the design elements from current literature on maturity model design) (Kitto et al. 2008). This approach is to some extent subjective, but multiple aspects guarantee objectivity or inter-subjectivity. First, textual data were official design documents (e.g. articles and websites), instead of subjective conversations, feelings or opinions. Lacity and Janson (1994) refer to a positivist (not interpretivist) text analysis, in which researchers are assumed to be outsiders who interpret texts from semantics without personal biases. Secondly, we collected multiple documents for the same BPMM, if possible. Thirdly, we acknowledged the importance of inter- and intrarater reliability. The author of this book was the main coder. In case of any confusion, other researchers were consulted to obtain a reliable coding and investigator triangulation. The texts were also repeatedly analysed over time, allowing refinements during the ongoing research, e.g. regarding the capability sub areas and after feedback from peers and experts.

1.7 Research Outline

After a first reading of our BPMM sample (N = 69), it turned out that slightly more than half of the models are non-academic. The vast majority is based on experiences and empirical research, e.g. case studies. Sometimes, BPMMs are preceded by a comparison of existing BPMMs, but (1) limited in the number of BPMMs and (2) limited to the content (i.e. which capability areas are measured on which lifecycle levels), instead of comparing all formal design elements (e.g. how capability areas are measured, etc.). Some BPMMs are also preceded by a theoretical study to identify the capability areas, but no consensus exists among the sampled BPMMs. Hence, the sample gave evidence for the existence of many BPMMs, but without a theoretical foundation. Therefore, our research has primarily a scientific contribution by summarising and grounding the BPMM literature to some degree, and by directing future BPMM research avenues. Its practical contribution mainly resides in the overview of BPMMs, and its advice on how to choose a BPMM that best fits the needs of particular organisations.

Figure 1.8 shows the main parts of our research with its corresponding outputs. Besides an introductory and concluding chapter, the book bundles three studies in Chap. 2.

1.7.1 Definitions

The first study contributes to the first and second issue of our problem statement by elaborating on the definitions for the main concepts in our research. A common understanding is appropriate, particularly because the collected BPMMs originate from different sources in which a different terminology is used.

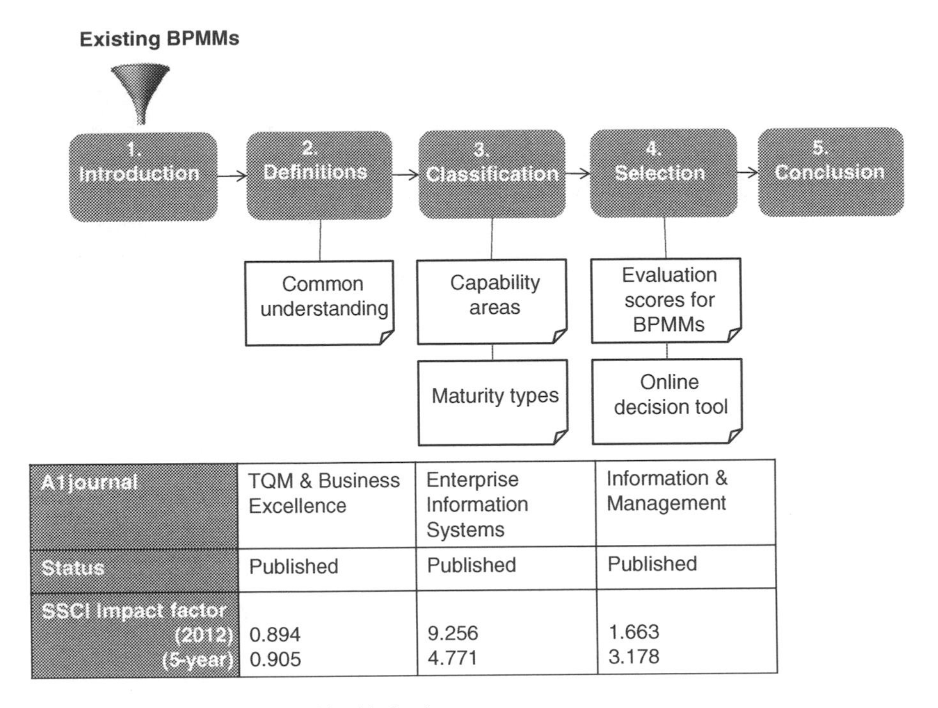

A1journal	TQM & Business Excellence	Enterprise Information Systems	Information & Management
Status	Published	Published	Published
SSCI Impact factor (2012) (5-year)	0.894 0.905	9.256 4.771	1.663 3.178

Fig. 1.8 The studies discussed in this book

- RQ1a. What is the appropriate BPMM scope? When does a maturity model consider business processes, and can be called a BPMM?
- RQ1b. What is the appropriate BPMM terminology? When does a maturity model consider maturity and capability?
- RQ1c. What is the appropriate BPMM design? When does a maturity model provide practical guidance to achieve process excellence?

Reference to the literature study:

Van Looy A, De Backer M, Poels G (2011) Defining business process maturity. A journey towards excellence. Total Qual Manag Bus Excell 22(11):1119–1137. doi:10.1080/14783363.2011.624779

1.7.2 Classification

The second study contributes to all issues of our problem statement by focusing on the capability areas that must be assessed and improved to obtain mature processes (i.e. which are capable to perform excellently). Capability areas differentiate a BPMM from other types of maturity models, and also have the potential to differentiate between BPMMs. Therefore, we first conducted a literature study on the critical success factors for business (process) excellence in order to provide a theoretical foundation of the capability areas. Afterwards, the collected BPMMs were mapped to the theoretically identified capability areas. By conducting a statistical classification (i.e. cluster analysis and discriminant analysis), we verified whether all BPMMs in the sample actually measure the same type of maturity, or whether different dimensions must be taken into account. Previous research already suggests the existence of two types of business process maturity, i.e. for the management of specific business processes and for the management of all business processes in the organisation (de Bruin and Rosemann 2007). This study evaluates whether the two dimensions are also present in our sample, and whether refinements can be added.

- RQ2a. Which capability areas can be assessed and improved by a BPMM in order to reach business (process) excellence?
- RQ2b. Which capability areas are actually assessed and improved by existing BPMMs?
- RQ2c. If RQ2b shows that different capability areas are actually assessed and improved, do existing BPMMs measure different types of maturity?

Reference to the classification study:

Van Looy A, De Backer M, Poels G (2012) A conceptual framework and classification of capability areas for business process maturity. Enterp Inf Syst. doi:10.1080/17517575.2012.688222

1.7.3 Selection

The third and final study within this book contributes to the first and second issue of our problem statement by extending our comparative study from capability areas to all aspects related to BPMMs. Although analysed, the descriptive statistics that describe our BPMM sample on those additional aspects are deliberately not detailed in this doctoral thesis. Instead, we concentrate on the most novel research results. For instance, the percentages of collected BPMMs that address the different variable options strongly depend on the sample, i.e. they change when more BPMMs will be collected. Such percentages give an impression on how BPMM characteristics are actually applied, without normative conclusions on how it should be or which trade-offs must be made. Consequently, study 3 examines which BPMMs must be chosen when. By content analysis, the main BPMM characteristics were identified as possible criteria that an organisation can consider when selecting a BPMM out of the wide array. Further on, an international Delphi study was conducted to derive a critical set of the most relevant decision criteria. The Delphi experts also assigned weights, which allowed calculating evaluation scores for existing BPMMs. To give guidance on how to select a BPMM that best fits the organisational needs, the resulting criteria were translated into an online questionnaire, called BPMM Smart-Selector. While filling out the questionnaire, a decision table navigates through the BPMM sample and suggests the best matching

BPMM. Hence, the expected classification of the second study is supplemented by critical BPMM considerations that motivate a rational BPMM choice.

- RQ3a. Which criteria help users (i.e. organisations or academics) choose a BPMM?

Reference to the selection study:
Van Looy A, De Backer M, Poels G, Snoeck M (2013). Choosing the right business process maturity model. Inf Manag. doi:10.1016/j.im.2013.06.002

1.8 Publications

We conclude this introductory chapter by listing the publications and conference contributions that currently resulted from this PhD research.

D1–Dissertation:

- Van Looy A (2012) Business process maturity. A comparative study on a sample of business process maturity models (Doctoral dissertation, Ghent University).

A1–Publications in international peer reviewed journals, available in the Web of Science:

- Van Looy A, De Backer M, Poels G (2011) Defining business process maturity. A journey towards excellence. Total Qual Manag Bus Excell 22(11):1119–1137. doi:10.1080/14783363.2011.624779
- Van Looy A, De Backer M, Poels G (2012) A conceptual framework and classification of capability areas for business process maturity. Enterp Inf Syst. doi:10.1080/17517575.2012.688222
- Van Looy A, De Backer M, Poels G, Snoeck M (2013) Choosing the right business process maturity model. Inf Manag. doi:10.1016/j.im.2013.06.002

A4—Publications in a journal not A2 or A3:

- Van Looy A (2010) Alweer een nieuw maturiteitsmodel? Het business process maturity model van OMG. Informatie 52(1):24–31
- Van Looy A (2013) Which business process maturity model best fits your organization? BPTrends, (July), 1.6

P1—Publications in international conference proceedings, available in the Web of Science:

- Van Looy A (2010) Does IT matter for business process maturity? A comparative study on business process maturity models. In: Meersman R et al. (eds). On the move to meaningful internet systems: OTM 2010 workshops, LNCS 6428. Springer, Berlin-Heidelberg, pp 687–697

→ ***Award for best contribution***

- Van Looy A, De Backer M, Poels G (2012) Towards a decision tool for choosing a business process maturity model. In: Peffers K, Rothenberger M, Kuechler B (eds) Design science research in information systems. Advances in theory and practice, LNCS 7286. Springer, Berlin-Heidelberg, pp 78–87
- Van Looy A (2012) Looking for a fit for purpose. Business process maturity models from a user's perspective. In: Poels G (ed) Enterprise information systems of the future. CONFENIS 2012, LNBIP 139. Springer, Berlin-Heidelberg, pp 182–189

C1—Other publications in international conference proceedings:

- Van Looy A, De Backer M, Poels G (2010) Which maturity is being measured? A classification of business process maturity models. In: Van Dongen BF, Reijers HA (eds) Proceedings of the fifth SIKS/BENAIS conference on enterprise information systems. Eindhoven University of Technology, Eindhoven, pp 7–16, 16 Nov 2010
- Van Looy A, De Backer M, Poels G (2011) Questioning the design of business process maturity models. In: Dignum V, Hidders J, Overbeek S (eds) Proceedings of the Sixth SIKS conference on enterprise information systems. Delft University of Technology, Delft, pp 51–60, 31 Oct 2011

Other conference and workshop contributions:

- Van Looy A (2011) Assessing and improving the maturity of cross-organisational business processes. Proceedings of the PhD symposium associated with the 9th international conference on business process management, Clermont-Ferrand, LIMOS Laboratory, CNRS, Université Blaise Pascal, France, 28 Aug 2011

FEB PhD Day contributions:

- Van Looy A (2010) Do we need a barometer for business process maturity within multi-level governance? (poster presentation). In: Third PhD Day. Ghent University, Ghent, 28 May 2010
- Van Looy A (2011) Business process maturity across organisations (poster presentation). In: Fourth PhD Day. Ghent University, Ghent, 24 May 2011
- Van Looy A (2012) Business process maturity (paper presentation). In: Fifth PhD Day. Ghent University, Ghent, 25 May 2012

FEB working papers:

- Van Looy A, De Backer M, Poels G (2011) Defining business process maturity. A journey towards excellence. FEB working paper 2011/725. Ghent University, Ghent

- Van Looy A, De Backer M, Poels G (2011) A theoretical framework and classification of capability areas for business process maturity. FEB working paper 2011/743. Ghent University, Ghent
- Van Looy A, De Backer M, Poels G (2012) Choosing the right business process maturity model. FEB working paper 2012/808. Ghent University, Ghent

Appendix: The Collected BPMMs (N = 69)

ID	Author(s)	BPMM name	Reference(s)
(1) Business process			
(1.1) Academic			
AOU	Aouad, Cooper, Hinks, Kagioglou and Sexton	Co-maturation model for synchronising BP and IT	(Aouad et al. 1998, 1999)
ARM	Armistead, Machin and Pritchard	BPM's degree of progress (as part of a larger survey)	(Armistead and Machin 1997; Pritchard and Armistead 1999)
DET	DeToro and McCabe	Process condition rating model	(DeToro and McCabe 1997)
HAM	Hammer	Process and enterprise maturity model (PEMM)	(Hammer 2007)
HAR1	Harrington	Process maturity grid	(Harrington 1991, 2006)
LEE	Lee, Lee and Kang	Value-based process maturity model (vPMM)	(Lee et al. 2007, 2009)
MAU	Maull, Tranfield and Maull	BPR maturity model	(Maull et al. 2003)
MCC1	McCormack and Johnson	BPO maturity model	(McCormack and Johnson 2001; McCormack 2007a, b)
ROH	Rohloff	Process management maturity assessment (PMMA)	(Rohloff 2009a, b)
ROS	Rosemann, de Bruin and Power	BPM maturity model	(de Bruin and Rosemann 2007; Hüffner 2004; Rosemann and de Bruin 2005a, b)
SEI	Software Engineering Institute (Carnegie Mellon University)	• Capability maturity model integration (CMMI) • Standard CMMI appraisal method for process improvement (SCAMPI)	(SEI 2000a, b, 2002a, b, 2006a, b, c, 2007, 2009)
SKR	Skrinjar, Bosilj-Vuksic, Stemberger and Hernaus	BPO maturity model	(Skrinjar et al. 2007, 2008)
WIL	Willaert, Van den Bergh, Willems and Deschoolmeester	Holistic BPO maturity framework	(Willaert et al. 2007)

(continued)

(continued)

ID	Author(s)	BPMM name	Reference(s)
(1.2) Non-academic			
BIS	Bisnez Management, Business and IT Trends Institute, students of Erasmus University, BPM magazine, information magazine	BPM maturity model (in Dutch: 'BPM volwassenheidsmodel')	(Bisnez Management 2010)
BPM	BPM Institute	State of BPM (part of a BPM survey)	(BPM Institute 2010; Dwyer 2006)
BPT	BP transformations group and BP group (previously BPM group)	8 Omega ORCA (Organisational readiness and capability assessment)	(BPT Group 2008; Towers 2009)
CAM1	CAM-I, Consortium for Advanced Management-International	Process-based management loop: • discipline model (organisation's current philosophy, business model, methods and tools) • process-based management assessment model (components) • process continuum model (levels)	(Dowdle et al. 2005)
CAM2	CAM-I, Consortium for Advanced Management-International	Process-based management assessment and implementation road map	(Dowdle and Stevens 2007)
CHA	Champlin (ABPMP)	Process management maturity model	(Champlin 2008)
DEL	Deloitte and Utrecht University	Business maturity model and scan	(Deloitte 2010)
ESI1	ESI, European Software Institute	EFQM/SPICE integrated model	(Ostolaza and Garcia 1999)
FAA	Federal aviation administration	• FAA integrated capability maturity model (FAA-iCMM) • FAA-iCMM appraisal method (FAM)	(FAA 2001, 2004, 2006)
FIS	Fisher (bearing point)	Business process maturity model	(Fisher 2004)
GAR1	Gardner	Process improvement road map	(Gardner 2004)
GAR2	Gartner	BPM maturity and adoption model	(Melenovsky and Sinur 2006)
HAR2	Harmon (BPTrends)	Informal BP maturity evaluation model	(Harmon 2004)

(continued)

(continued)

ID	Author(s)	BPMM name	Reference(s)
IDS	IDS scheer, software AG	• BPM maturity check • BPM road map assessment	(IDS Scheer 2010; Luyckx 2007; Scheer and Brabaender 2009)
ISO	ISO/IEC, International Organisation for Standardisation and International Electrotechnical Commission	ISO/IEC 15504	(ISO/IEC 2003, 2004a, b, c, d, 2008, ISOSPICE 2010; SQI 2007)
O&I	O&i	BPM scan	(O&i 2010; Tolsma and de Wit 2009)
OMG	Object management group	BPMM	(OMG 2008)
ORA	Oracle and BEA systems	BPM lifecycle assessment survey	(Oracle 2008a, b)
REM	Remoreras	Process culture maturity model	(Remoreras 2009)
RUM	Rummler-Brache group	Process performance index	(Rummler-Brache Group 2004)
SAP	SAP	Process maturity analysis and plan	(Scavillo 2008)
SCH1	Scheer	BPM check-up	(Scheer 2007)
SMI	Smith and Fingar	Process management maturity model (PMMM)	(Smith and Fingar 2004)
SPA	Spanyi	BP competence grid	(Spanyi 2004a; b)
(2) Supply chain			
(2.1) Academic			
ARY	Aryee, Naim and Lalwani	SC integration maturity model	(Aryee et al. 2008)
BOH	Böhme and Childerhouse	SC integration evaluation tool and maturity model	(Böhme 2008)
CAM3	Campbell and Sankaran	SC integration enhancement framework (SCIEF)	(Campbell and Sankaran 2005)
MCC2	McCormack et al.	SC management maturity model	(Lockamy III and McCormack 2004, 2007; McCormack 2007c, 2008)
MCL	McLaren	Web-enabled SC integration measurement model	(McClaren 2006)
MIC	Michigan State University	Twenty-first century logistics framework	(Closs and Mollenkopf 2004; Mollenkopf and Dapiran 2005)
NET	Netland, Alfnes and Fauske	SC maturity assessment test (SCMAT)	(Netland et al. 2007; Netland and Alfnes 2008)
RIV	Riverola	SC management—technology maturity model	(Riverola 2001)

(continued)

(continued)

ID	Author(s)	BPMM name	Reference(s)
TOK	Tokyo Institute of Technology	Logistics scorecard (LSC)	(Enkawa 2005; Yaibuathet et al. 2007, 2008)
(2.2) Non-academic			
ABE	Aberdeen group	Global SC maturity framework	(AberdeenGroup 2006)
AND	Andersen consulting (accenture)	SC continuum	(Anderson and Lee 2000)
CGF	CGF, consumer goods forum (former GCI, global commerce initiative)	Global scorecard for efficient consumer response capability	(CGF 2010)
CGR	CGR management consulting	SC management maturity model	(Ayers 2010a, b)
CHI	Chicago consulting	SC maturity model	(Chicago Consulting 2010)
CSC	CSC, SC management review magazine and Michigan State University	• SC maturity model (until 2006) • Ten SC competencies (as from 2007)	(Poirier et al. 2010)
EKN	eKNOWtion	SC maturity monitor (SCM^2)	(eKNOWtion 2009)
IBM	IBM	SC maturity model	(IBM 2007)
JER	Jeroen van den Bergh Consulting and VU University Amsterdam	SC maturity scan	(van den Bergh 2010)
LMI	LMI Research Institute	GAIA SC sustainability maturity model	(Boone et al. 2009)
MAN	Manugistics and JDA software	SC compass	(Holmes 1997)
PMG	PMG and PRTM	SC maturity model	(Cohen and Roussel 2005)
SCC	SCC, supply chain council and APQC	SCORmark survey (for benchmarking, resulting in an improvement road map)	(APQC 2007; Supply-Chain Council 2008)
SCH2	Schoenfeldt	SC mgt maturity model	(Schoenfeldt 2008)
STE	Stevens	SC integration model	(Stevens 1989)
(3) Collaboration			
(3.1) Academic			
FRA	Fraser, Farrukh and Gregory	Collaboration maturity grid (for new product introduction and development)	(Fraser et al. 2003)
MAG	Magdaleno, Cappelli, Baiao, Santoro and Araujo	Collaboration maturity model (ColabMM)	(Magdaleno et al. 2008)
RAM	Ramasubbu and Krishnan	Process maturity framework for managing distributed software product development	(Ramasubbu and Krishnan 2005)

(continued)

(continued)

ID	Author(s)	BPMM name	Reference(s)
SIM	Simatupang and Sridharan	SC Collaboration index	(Simatupang and Sridharan 2005)
TAP	Tapia, Daneva, van Eck and Wieringa	IT-enabled collaborative networked organisations maturity model (ICoNOs MM)	(Tapia et al. 2008)
WOG	Wognum and Faber	Fast reactive extended enterprise—capability assessment framework (FREE-CAF)	(Wognum and Faber 2002)
(3.2) Non-academic			
ESI2	European Software Institute	Enterprise collaboration maturity model	(de Soria et al. 2009)
VIC	Voluntary interindustry commerce standards	Collaborative planning, forecasting and replenishment (CPFR) rollout readiness self-assessment	(VICS 2004)

References

Aberdeen Group (2006) Global supply chain benchmark report. Industry priorities for visibility, B2B collaboration, trade compliance, and risk management. https://www-935.ibm.com/services/us/igs/pdf/aberdeen-benchmark-report.pdf. Accessed 23 June 2010

Ahern DM, Clouse A, Turner R (2004) CMMI distilled. A practical introduction to integrated process improvement. SEI series in software engineering, 2nd edn. Pearson Education, Boston

Alden JW (2007) Measuring process maturity: the business process maturity model. On the business transformation conference, Transformation & innovation 2007. http://www.slideshare.net/TransformationInnovation/measuring-process-maturity-the-business-process-maturity-model. Accessed 2 Dec 2009

Anderson DL, Lee HL (2000) The internet-enabled supply chain: from the "first click" to the "last mile". In: Anderson DL (ed) Achieving supply chain excellence through technology, vol 2. Montgomery Research, pp 15–20

Aouad G, Cooper R, Kagioglou M, Hinks J, Sexton M (1998) A synchronised process/IT model to support the co-maturation of processes and IT in the construction sector (CIB report). Time Research Institute, Salford

Aouad G, Kagioglou M, Cooper R, Hinks J, Sexton M (1999) Technology management of IT in construction: a driver or an enabler? Logist Inf Manag 12(1/2):130–137

APQC (2007) Supply-chain council SCOR-mark survey. http://www.apqc.org/scc. Accessed 23 April 2010

Armistead C, Machin S (1997) Implications of business process management for operations management. Int J Oper Prod Manag 17(9):886–898

Aryee G, Naim MM, Lalwani C (2008) Supply chain integration using a maturity scale. J Manufact Tech Manag 19(5):559–575

Attaran M (2004) Exploring the relationship between information technology and business process reengineering. Inf Manag (41):585–596

Attaran M, Attaran S (2004) The rebirth of re-engineering: X-engineering. Bus Process Manag J 10(4):415–429
Ayers JB (2010a) Supply chain maturity self-assessment. http://ayers-consulting.com/SC%20Maturity%20Self%20Assessment.htm. Accessed 23 June 2010
Ayers JB (2010b) Supply chain project management. A structured collaborative and measurable approach, 2nd edn. Taylor and Francis Group, Boca Raton
Basu SC, Palvia PC (2000) Business process reengineering. In: Kent A (ed) Encyclopedia of library and information science, vol 67. Marcel Dekker, New York, pp 24–34
Bauer B, Müller JP, Roser S (2006) A decentralized broker architecture for collaborative business process modelling and enactment. Proceedings of the 2nd international conference on interoperability for enterprise software and applications, Bordeaux. Springer, London, pp 115–125
Becker J, Knackstedt R, Pöppelbuss J (2009) Developing maturity models for IT management. A procedure model and its application. Bus Inf Syst Eng 1(3):213–222
Becker J, Niehaves B, Pöppelbuss J, Simons A (2010) Maturity models in IS research. Proceedings of the 18th European conference on information systems, AIS Electronic Library, Pretoria, pp 1–12
Beckford J (2009) Quality. A critical introduction, 3rd edn. Routledge, London
Bisnez Management (2010) Business process management onderzoek 2009–2010. http://www.bisnez.org/files/Rapport_BPM_volwassenheid_onderzoek_2009_2010_def.pdf. Accessed 20 June 2010
Böhme T (2008) Supply chain integration: a case-based investigation of status, barriers, and paths to enhancement. The University of Waikato, Waikato
Boone LM, Colaianni AJ, Hardison JR, Shafer JJ, Shepherd NJ, Ramaswamy MS et al (2009) The GAIA supply chain sustainability maturity model (Report IR927R1). http://www.lmi.org/Logistics/Documents/GAIA_Sustainable_Supply_Chain_Maturity_Model.pdf. Accessed 23 June 2010
BPM Institute (2010) 2010 BPM market assessment survey. http://2010stateofbpm.surveyconsole.com/. Accessed 20 June 2010
BPT Group (2008) Welcome to 8 Omega. Version 2.0. http://bptg.seniordev.co.uk/8omega.aspx. Accessed 20 June 2010
Buglione L (2009) Software engineering improvement. http://www.semq.eu/leng/proimpsw.htm. Accessed 25 Feb 2010
Burlton R (2010) Delivering business strategy through process management. In: vom Brocke J, Rosemann M (eds), Handbook on business process management, vol 2. Springer, Berlin Heidelberg, pp 5–37
Burlton TR (2001) Business process management: profiting from processes. Sams Publishing, Indianapolis
Campbell J, Sankaran J (2005) An inductive framework for enhancing supply chain integration. Int J Prod Res 43(16):3321–3351
Carr NG (2003) IT doesn't matter. Harv Bus Rev 81(5):41–49
CGF (2010) The global scorecard. http://www.globalscorecard.net/. Accessed 23 June 2010
Champlin B (2008) Dimensions of business process change. https://www.bpminstitute.org/uploads/media/Champlin-6-25-08.pdf. Accessed 20 June 2010
Champy J (2002) X-engineering the corporation. Reinventing your business in the digital age. Warner Business Books, New York
Chang JF (2006) Business process management systems. Strategy and implementation. Taylor & Francis Group, Boca Raton
Chicago Consulting (2010) Supply chain maturity. A self-assessment. http://www.chicago-consulting.com/supplyChainMaturityAssessment.shtml. Accessed 23 June 2010
Closs DJ, Mollenkopf DA (2004) A global supply chain framework. Ind Mark Manage 33:37–44
Cobb CG (2003) From quality to business excellence. A systems approach to management. ASQ Quality Press, Wisconsin

Cohen S, Roussel J (2005) Strategic supply chain management. The 5 disciplines for top performance. McGraw-Hill, New York
Coombs R, Hull R (2001) The politics of IT strategy and development in organizations. In: Dutton WH (ed) Information and communication technologies. Oxford University Press, New York, pp 159–176
Crosby PB (1979) Quality is free. The art of making quality certain. McGraw-Hill, New York
Curtis B, Alden J (2007) The business process maturity model: what, why and how. http://www.bptrends.com/publicationfiles/02-07-COL-BPMMWhatWhyHow-CurtisAlden-Final.pdf. Accessed 2 Dec 2009
Curtis B, Hefley WE, Miller SA (2001) The people capability maturity model. Guidelines for improving the workforce. SEI series management of human resources. Pearson Education, Boston
Curtis B, Kellner MI, Over J (1992) Process modeling. Commun ACM 35(9):75–90
Davenport TH (1993) Process innovation. Reengineering work through information technology. Harvard Business School, Boston
de Bruin T, Rosemann M (2007) Using the Delphi technique to identify BPM capability areas. Proceedings of the 18th Australasian conference on information systems, Toowoomba, pp 642–653, 5–7 Dec 2007
de Bruin T, Freeze R, Kulkarni U, Rosemann M (2005) Understanding the main phases of developing a maturity assessment model. Proceedings of the 16th Australasian conference on information systems. AISeL, Sydney, p 10, 29 Nov-2 Dec
de Soria IM, Alonso J, Orue-Echevarria L, Vergara M (2009) Developing an enterprise collaboration maturity model: research challenges and future directions. Proceeding of the 15th international conference on concurrent enterprising. ICE, Leiden, 22–24 June 2009
Debevoise T (2007) Converging BPM and business rules maturity models. http://www.bpminstitute.org/articles/article/article/converging-bpm-and-business-rules-maturity-models.html. Accessed 23 June 2010
Deloitte (2010) Het business maturity model. http://www.deloitte.com/view/nl_NL/nl/diensten/consulting/strategy-operations/business-maturity-model/index.htm. Accessed 21 June 2010
Delphi Group (2002) BPM2002 market milestone report. White paper. Delphi Group, Boston
Deming WE (1994) The new economics: for industry, government, education, 2nd edn. Center for Advanced Educational Services, Cambridge
Deming WE (2000) Out of the crisis (originally published in 1982), 3rd edn. Massachusetts Institute of Technology Press, Cambridge
DeToro I, McCabe T (1997) How to stay flexible and elude fads. Qual Prog:55–60. March
Dowdle P, Stevens J (2007) The process audit and PBM roadmap. A PBM program perspective. http://www.cam-i.org/docs/PBM_and_PEMM.pdf. Accessed 21 June 2010
Dowdle P, Stevens J, McCarty B, Daly D (2005) The process-based management loop. J Corp Account Finance:55–60. Jan
Dwyer T (2006) BPMInstitute's state of business process management. Assessing the current state of BPM awareness and usage. An executive white paper. http://www.bpminstitute.org/uploads/media/2006_State_of_BPM_-_EOY.pdf. Accessed 20 June 2010
Edwards C, Braganza A, Lambert R (2000) Understanding and managing process initiatives: A framework for developing consensus. Knowl Proc Manag 7(1):29–36
EFQM (2010) EFQM—the official website. http://www.efqm.org. Accessed 11 Jan 2010
eKNOWtion (2009) Supply chain maturity monitor. http://www.eknowtion.com/phpQ/fillsurvey.php?sid=8. Accessed 23 June 2010
El Emam K, Birk A (2000) Validating the ISO/IEC 15504 measure of software requirements analysis process capability. IEEE Trans Software Eng 26(6):541–566
Enkawa T (2005) Logistics scorecard (LSC) in Japan. Logistics networks are successfully benchmarked in Japan? http://transportal.fi/Hankkeet/eglo/.383.pdf. Accessed 23 June 2010
FAA (2001) FAA-iCMM. Version 2.0. An integrated capability maturity model for enterprise-wide improvement. http://www.faa.gov/about/office_org/headquarters_offices/aio/library/. Accessed 15 Mar 2010

FAA (2004) Safety and security extensions for integrated capability maturity models. http://www.faa.gov/about/office_org/headquarters_offices/aio/library/. Accessed 15 Mar 2010

FAA (2006) FAA-iCMM appraisal method (FAM). Version 2.0. http://www.faa.gov/about/office_org/headquarters_offices/aio/library/. Accessed 15 Mar 2010

Fisher DM (2004) The business process maturity model. A practical approach for identifying opportunities for optimization. http://www.bptrends.com/publicationfiles/10-04%20ART%20BP%20Maturity%20Model%20-%20Fisher.pdf. Accessed 23 June 2010

Foggin JH, Mentzer JT, Monroe CL (2004) A supply chain diagnostic tool. Int J Phys Distrib Logist Manag 34(10):827–855

Fox C, Frakes W (1997) The quality approach: is it delivering? Commun Assoc Comput Mach 40(6):24–29

Fraser P, Farrukh C, Gregory M (2003) Managing product development collaborations—a process maturity approach. Proc Inst Mech Eng Part B: J Eng Manuf 217(11):1499–1519

Frohlich MT, Westbrook R (2001) Arcs of integration: an international study of supply chain strategies. J Oper Manag 19:185–200

Gardner RA (2004) The process-focused organization. A transition strategy for success. ASQ, Quality Press, Milwaukee

Gartner (2012) Gartner executive program's worldwide survey of more than 2,300 CIOs shows flat IT budgets in 2012, but IT organizations must deliver on multiple priorities. http://www.gartner.com/it/page.jsp?id=1897514. Accessed 26 July 2012

Glaser BG, Strauss AL (1967) The discovery of grounded theory: strategies for qualitative research. Transaction Publishers, New Jersey

Goethals F (2006) Classifying and assessing extended enterprise integration approaches. K.U.Leuven, Leuven

Gregor S (2006) The nature of theory in information systems. MIS Q 30(3):611–642

Hain S (2010) Developing a situational maturity model for collaboration (SiMMCo)—measuring organizational readiness. Proceedings of the fifth international conference DESTRIST. St. Gallen, DESTRIST, p 6, 4–5 June 2010

Hammer M (1990) Reengineering work: don't automate, obliterate. Harv Bus Rev 68(4):104–112

Hammer M (1996) Beyond reengineering. HarperCollins Publishers, New York

Hammer M (2007) The process audit. Harv Bus Rev 4:111–123

Hammer M, Champy J (2003) Reengineering the corporation. A manifesto for business revolution (originally published in 1993), 2nd edn. HarperCollins Publishers, New York

Hammer M, Stanton SA (1995) The reengineering revolution. A handbook. HarperCollins Publishers, New York

Harmon P (2004) Evaluating an organization's business process maturity. http://www.bptrends.com/publicationfiles/03-04%20NL%20Eval%20BP%20Maturity%20-%20Harmon.pdf. Accessed 23 June 2010

Harmon P (2009). Process maturity models. http://www.bptrends.com/publicationfiles/spotlight_051909.pdf. Accessed 24 Nov 2010

Harmon P (2010) Business process frameworks. http://www.bptrends.com/publicationfiles/spotlight_03232010.pdf. Accessed 24 Nov 2010

Harrington HJ (1991) Business process improvement. The breakthrough strategy for total quality, productivity, and competitiveness. McGraw-Hill, New York

Harrington HJ (2006) Process management excellence. The art of excelling in process management. Book 1 in the five-part series. The five pillars of organizational excellence. Paton Press, California

Harrington HJ, Harrington JS (1995) Total improvement management. The next generation in performance improvement. McGraw-Hill, New York

Harrington HJ, Esseling EK, Van Nimwegen H (1997) Business process improvement workbook. Documentation, analysis, design, and management of business process improvement. McGraw-Hill, New York

Harter DE, Slaughter SA (2000) Process maturity and software quality: a field study. Proceedings of the twenty-first international conference on information systems. Association for Information Systems, Brisbane, pp 407–411

Hevner AR, March ST, Park J, Ram S (2004) Design science in information systems research. MIS Q 28(1):75–105

Holmes J (1997) Building supply chain communities. http://criticalcomputing.com/. Accessed 24 April 2010

Holweg M, Disney S, Holmström J, Smaros J (2005) Supply chain collaboration: making sense of the strategy continuum. Europ Manag J 23(2):170–181

Houy C, Fettke P, Loos P (2010) Empirical research in business process management—analysis of an emerging field of research. Bus Proc Manag J 16(4):619–661

Hüffner T (2004) The BPM maturity model—towards a framework for assessing the business process management maturity of organisations. GRIN, Munich

Humphrey WS (1989) Managing the software process. SEI series in software engineering. Addison-Wesley, Boston

IBM (2007) The IBM global business services 2007 Mainland China value chain study. http://www-935.ibm.com/services/us/index.wss/ibvstudy/gbs/a1028901?cntxt=a1005268#additional_resources. Accessed 23 June 2010

Ibrahim L, Pyster A (2004) A single model of process improvement. Lessons learned at the US federal aviation administration. IT Pro—IEEE Comp Soc:43–49. May-July

IDS Scheer (2010) BPM maturity check. http://www.bpmmaturity.com/. Accessed 22 June 2010

ISO (2009) International Organization for Standardization—the official website. http://www.iso.org/. Accessed 6 Jan 2010

ISO/IEC (2003) Software engineering—process assessment—part 2: performing an assessment—ISO/IEC 15504-2:2003(E). ISO/IEC, Geneva

ISO/IEC (2004a) Information technology—process assessment—part 1: concepts and vocabulary—ISO/IEC 15504-1:2004(E). ISO/IEC, Geneva

ISO/IEC (2004b) Information technology—process assessment—part 3: guidance on performing an assessment—ISO/IEC 15504-3:2004(E). ISO/IEC, Geneva

ISO/IEC (2004c) Information technology—process assessment—part 4: guidance on use for process improvement and process capability determination—ISO/IEC 15504-4:2004(E). ISO/IEC, Geneva

ISO/IEC (2004d) Software engineering—process assessment—part 2: performing an assessment—technical corrigendum 1—ISO/IEC 15504-2:2003/Cor.1:2004(E). ISO/IEC, Geneva

ISO/IEC (2008) Information technology—process assessment—part 7: assessment of organizational maturity—ISO/IEC TR 15504-7:2008(E). ISO/IEC, Geneva

ISOSPICE (2010) ISOSPICE. http://www.isospice.com/. Accessed 26 Feb 2010

IT Governance Institute (2007) COBIT 4.1—framework, control objectives, management guidelines, maturity models. IT Governance Institute, Rolling Meadows

Jablonski S, Bussler C (1996) Workflow management: modeling concepts, architecture, and implementation. Int Thomson Comp Press, London

Jiang JJ, Klein G, Hwang H-G, Huang J, Hung S-Y (2004) An exploration of the relationship between software development process maturity and project performance. Inf Manag 41:279–288

Kannengiesser U (2008) Subsuming the BPM life cycle in an ontological framework of designing. Proceedings of the CIAO! and EOMAS workshops, CAiSE. Springer, Berlin Heidelberg, pp 31–45

Kaplan RS, Norton DP (2001) The strategy-focused organization. How balanced scorecard companies thrive in the new business environment. Harvard Business School Press, Boston

Kellner MI, Madachy RJ, Raffo DM (1999) Software process simulation modeling: why? what? how? J Syst Softw 46(2/3):1–18

Khalfan MM, Anumba CJ, Carrillo PM (2001) Development of a readiness assessment model for concurrent engineering in construction. Benchmarking: Int J 8(3):223–239

Kitto SC, Chesters J, Grbic C (2008) Quality in qualitative research. Criteria for authors and assessors in the submission and assessment of qualitative research articles for the Medical Journal of Australia. Med J Aust 188(4):243–246

Kulpa MK, Johnson KA (2008) Interpreting the CMMI. A process improvement approach, 2nd edn. Taylor & Francis Group, Boca Raton

Kwak YH, Ibbs CW (2002) Project management process maturity (PM^2) model. J Manag Eng 7:150–155

Lacity MC, Janson MA (1994) Understanding qualitative data: a framework of text analysis methods. J Manag Inf Syst 11(2):137–155

Lee J, Lee D, Kang S (2007) An overview of the business process maturity model (BPMM). In: Chang KC, Wang W, Chen L, Ellis CA, Hsu C, Tsoi AC et al Advances in web and network technologies, and information management APWeb/WAIM 2007 international workshops: DBMAN 2007, WebETrends 2007, PAIS 2007 and ASWAN 2007, Huang Shan, Chine. Proceedings. LNCS 4537, Springer, Berlin Heidelberg, pp 384–395, 16–18 June 2007

Lee J, Lee D, Kang S (2009) vPMM: a value based process maturity model. In: Lee R, Hu G, Miao H (eds) Computer and information science 2009. Springer, Berlin Heidelberg, pp 193–213

Lockamy A III, McCormack K (2004) The development of a supply chain management process maturity model using the concepts of business process orientation. Supply Chain Manag: Int J 9(4):272–278

Lockamy A III, McCormack K (2007) Supply chain maturity and performance. In: McCormack K (ed) Business process maturity. Theory and application. Booksurge Publishing, South Carolina, pp 105–135

Luftman J (2004) Assessing business-IT alignment maturity. In: Van Grembergen W (ed) Strategies for information technology governance. Idea Group Publishing, Hershey, pp 99–128

Luyckx F (2007) IDS BPM roadmap assessment. http://www.ids-scheer.nl/set/2238/Overview%20BPM%20maturity%20model.pdf. Accessed 22 June 2010

Magdaleno AM, Cappelli C, Baiao FA, Santoro FM, Araujo R (2008) Towards collaboration maturity in business processes: an exploratory study in oil production processes. Inf Syst Manag 25:302–318

Maier AM, Moultrie J, Clarkson JP (2009) Developing maturity grids for assessing organisational capabilities: practitioner guidance. Proceedings of the 4th international conference on management consulting, academy of management (MCD'09), 11–13 June 2008, Vienna, Austria. Management Consulting Division, Vienna, p 29

Maier AM, Moultrie J, Clarkson PJ (2012) Assessing organizational capabilities: reviewing and guiding the development of maturity grids. IEEE Trans Eng Manage 59(1):138–159

March ST, Smith GF (1995) Design and natural science research on information technology. Decis Support Syst 15(4):251–266

Maull RS, Tranfield DR, Maull W (2003) Factors characterising the maturity of BPR programmes. Int J Oper Prod Manag 23(6):596–624

McClaren T (2006) A measurement model for web-enabled supply chain integration. Paper 18. Proceedings of the 19th bled econference evalues, Bled, pp 1–13, 5–7 June 2006

McCormack K (2007a) BPO and business process maturity. In: McCormack K (ed) Business process maturity. Theory and application. Booksurge Publishing, South Carolina, pp 61–72

McCormack K (2007b) Introduction to the theory of business process orientation. In: McCormack K (ed) Business process maturity. Theory and application. Booksurge Publishing, South Carolina, pp 1–18

McCormack K (2007c) Supply chain management maturity. In: McCormack K (ed) Business process maturity. Theory and application. Booksurge Publishing, South Carolina, pp 73–103

McCormack K, Johnson WC (2001) Business process orientation: gaining the e-business competitive advantage. St. Lucie Press, Florida

McCormack K, Ladeira MB, de Oliveira MP (2008) Supply chain maturity and performance in Brazil. Supply Chain Manag Int J 13(4):272–282

McGovern J, Ambler SW, Stevens ME, Linn J, Sharan V, Jo EK (2004) A practical guide to enterprise architecture. Pearson Education, New Jersey

McKinsey (2011) A rising role for IT: McKinsey global survey results. https://www.mckinseyquarterly.com/High_Tech/Strategy_Analysis/A_rising_role_for_IT_McKinsey_Global_Survey_results_2900. Accessed 26 July 2012

Melenovsky MJ, Sinur J (2006) BPM maturity model identifies six phases for successful BPM adoption. Gartner Research, Stamford

Mettler T, Rohner P (2009) Situational maturity models as instrumental artifacts for organizational design. Proceedings of the 4th international conference on DESRIST, Malvern, p 9, 7–8 May 2009

Mollenkopf D, Dapiran GP (2005) The importance of developing logistics competencies: a study of Australian and New Zealand firms. Int J Logist: Res Appl 8(1):1–14

Moore JW (1999) An integrated collection of software engineering standards. IEEE Softw:51–57

Moultrie J, Clarkson PJ, Probert D (2007) Development of a design audit tool for SMEs. J Prod Innov Manag 24:335–368

Nave D (2002) How to compare Six Sigma, lean and the theory of constraints. A framework for choosing what's best for your organization. Qual Prog:73–78

Netland TH, Alfnes E, Fauske H (2007) How mature is your supply chain? Supply chain maturity assessment test. Proceedings of the 14th EurOMA conference, managing operations in an expanding Europe, Ankara, Turkey. EurOMA, Bilkent University, Ankara, pp 1–10, 17–20 June 2007

Netland T, Alfnes E (2008) A practical tool for supply chain improvement—experiences with the supply chain maturity assessment test (SCMAT). Proceedings of the 3rd world conference on production and operations management, Tokyo. Gakushuin University, Tokyo, pp 1–14, 5–8 Aug 2008

Nolan RL (1973) Managing the computer resource. A stage hypothesis. Commun ACM 16(7):399–405

O&i (2010) Doe nu de online BPM-scan. http://www.oi.nl/bpmscan/. Accessed 23 June 2010

OGC (2007) The official introduction to the ITIL service lifecycle. TSO—The Stationary Office, London

OGC (2010) Prince2. http://www.ogc.gov.uk/methods_prince_2__overview.asp. Accessed 11 May 2010

OMG (2008) Business process maturity model (BPMM). Version 1.0. http://www.omg.org/spec/BPMM/1.0/PDF. Accessed 2 Dec 2009

O'Neill P, Sohal AS (1999) Business process reengineering. A review of recent literature. Technovation 19:571–581

Oracle (2008a) BPM lifecycle assessment. http://bpmready.nvishweb.com/. Accessed 23 June 2010

Oracle (2008b) State of the business process management market 2008. An Oracle white paper. http://www.oracle.com/technologies/bpm/docs/state-of-bpm-market-whitepaper.pdf. Accessed 4 May 2010

Ostolaza E, Garcia AB (1999) EFQM/SPICE integrated model. The business excellence road for software intensive organizations. Proceedings of the international conference on product focused software improvement, Oulu, Finland. University of Oulu, VTT Electronics, Oulu, pp 437–452, 22–24 June 1999

Paulk MC (2004) Surviving the quagmire of process models, integrated models, and standards. Proceedings of the American society for quality annual quality conference, Toronto, Canada, p 8. AIS Electronic Library, ASQ, Toronto, 24–27 May 2004

Paulk MC (2008) A taxonomy for improvement frameworks. Proceedings of the world congress for improvement frameworks. Bethesda, Maryland, p 15, 17 Sept 2008

Paulk MC, Weber CV, Curtis B, Chrissis MB (1995) The capability maturity model. Guidelines for improving the software process. SEI series in software engineering. Addison Wesleym, Boston

Peffers K, Tuunanen T, Rothenberger MA, Chatterjee S (2007) A design science research methodology for information systems research. J Manag Inf Syst 24(3):45–77

Pesic AM (2009) Business process management maturity model and Six Sigma: An integrated approach for easier networking. Proceedings of the international conference on economics and management of networks, EMNet. Springer, Sarajevo, p 19, 3–5 Sept 2009

Plattfaut R, Niehaves B, Pöppelbuss J, Becker J (2011) Development of BPM capabilities—Is maturity the right path? Proceedings of 19th European conference on information systems. AIS Electronic Library, Helsinki, pp 1–12

Plomp MG, Batenburg RS (2010) Measuring chain digitisation maturity: an assessment of Dutch retail branches. Supply Chain Manag: Int J 15(3):227–237

Poirier CC, Quinn FJ, Swink ML (2010) Diagnosing greatness. Ten traits of the best supply chains. J. Ross Publishing, Fort Lauderdale

Pöppelbuss J, Röglinger M (2011) What makes a useful maturity model? A framework of general design principles for maturity models and its demonstration in business process management. Proceedings of the 19th European conference on information systems. AIS Electronic Library, Helsinki, pp 1–12

Porter ME (1985) Competitive advantage. Creating and sustaining superior performance. The Free Press, New York

Pritchard J-P, Armistead C (1999) Business process management—lessons from European business. Bus Proc Manag J 5(1):10–32

Ramasubbu N, Krishnan MS (2005) Leveraging global resources: A process maturity framework for managing distributed software product development. Paper 386. Proceedings of AMCIS, Omaha, Nebraska, USA. AIS Electronic Library, University of Nebraska, Omaha, pp 3085–3090, 11–14 Aug 2005

Ravesteyn P, Jansen S (2009) A situational implementation method for business process management systems. Proceedings of the fifth Americas conference on information systems. AISeL, San Francisco, p 9, 6–9 Aug 2009

Remoreras G (2009) Achieving the highest level of process culture maturity. http://mysimpleprocesses.com. Accessed 1 June 2010

Renken J (2004) Developing an IS/ICT management capability maturity framework. Proceedings of the 2004 annual research conference of the South African Institute of computer scientists and information technologists on IT research in developing countries, Stellenbosch, Western Cape, South Africa. SAICSIT, Stellenbosch, pp 53–62

Riverola J (2001) A general approach to the case gathering phase. http://catalcatel.iese.edu/WEBSCM/Archivos/MaturityModel.pdf. Accessed 6 July 2010

Röglinger M, Pöppelbuss J, Becker J (2012) Maturity models in business process management. Bus Proc Manag J 18(2):328–346

Rohloff M (2009a) Case study and maturity model for business process management implementation. In: Dayal U, Eder J, Koehler J, Reijers HA (eds) Business process management. 7th international conference, BPM 2009, Ulm, Germany. Proceedings. LNCS 5701, pp 128–142, Springer, Berlin Heidelberg, 8–10 Sept 2009

Rohloff M (2009b) Process management maturity assessment. Paper 631. Proceedings of the 15th Americas conference on information systems, San Fransisco, California. AISeL, San Fransisco, p 12, 6–9 Aug 2009

Rosemann M (2010) The service portfolio of a BPM center of excellence. In: vom Brocke J, Rosemann M (eds) Handbook on business process management 2. Springer, Berlin Heidelberg, pp 267–284

Rosemann M, de Bruin T (2005a) Application of a holistic model for determining BPM maturity. http://www.bptrends.com/publicationfiles/02-05%20WP%20Application%20of%20a%20Holistic%20Model-%20Rosemann-Bruin%20-%E2%80%A6.pdf. Accessed 8 Feb 2010

Rosemann M, de Bruin T (2005b) Towards a business process management maturity model. Proceedings of the 13th European conference on information systems, Regensburg, Germany. ECIS, Regensburg, p 12, 26–28 May 2005

Rosemann M, vom Brocke J (2010) The six core elements of business process management. In: vom Brocke J, Rosemann M (eds) Handbook on business process management, vol 1. Springer, Berlin Heidelberg, pp 107–122

Ross JE, Omachon VK (2004) Principles of total quality, 3rd edn. CRC Press, Florida

Rout TP, El Emam K, Fusani M, Goldenson D, Jung H-W (2007) SPICE in retrospect: developing a standard for process assessment. J Syst Softw 80:1483–1493

Rummler-Brache Group (2004) Business process management in U.S. firms today. http://rummler-brache.com/upload/files/PPI_Research_Results.pdf. Accessed 23 June 2010

Santanen E, Kolfschoten G, Golla K (2006) The collaboration engineering maturity model. Proceedings of the 39th Hawaii international conference on system sciences, vol 1. IEEE Computer Society, Los Alamitos, p 16c

Scavillo M (2008) Business process transformation in the software industry. http://www.sdn.sap.com/irj/scn/go/portal/prtroot/docs/library/uuid/70559771-c266-2b10-1499-8c36e668e0a6?QuickLink=events&overridelayout=true. Accessed 23 June 2010

Scheer A-W (2007) BPM = business process management = business performance management. http://www.professor-scheer-bpm.com/BPM_Scheer_Business_Process_Management_en.pdf. Accessed 27 Apr 2010

Scheer A-W, Brabaender E (2009) BPM governance. The process of business process management. ARIS expert paper. IDS Scheer AG, Saarbruecken

Schoenfeldt TI (2008) A practical application of supply chain management principles. ASQ, Milwaukee

Scott JE (2007) Mobility, business process management, software sourcing, and maturity model trends: propositions for the IS Organization of the future. Inf Syst Manag 24:139–145

SEI (1987) A method for assessing the software engineering capability of contractors. http://www.sei.cmu.edu/reports/87tr023.pdf. Accessed 22 Feb 2010

SEI (1993) Capability maturity model for software. Version 1.1. http://www.sei.cmu.edu/reports/93tr024.pdf. Accessed 22 Feb 2010

SEI (2000a) CMMI for systems engineering/software engineering. Version 1.02—continuous representation. http://www.sei.cmu.edu/reports/00tr019.pdf. Accessed 11 Feb 2010

SEI (2000b) CMMI for systems engineering/software engineering. Version 1.02—staged representation. http://www.sei.cmu.edu/reports/00tr018.pdf. Accessed 11 Feb 2010

SEI (2002a) CMMI for systems engineering, software engineering, integrated product and process development, and supplier sourcing. Version 1.1—continuous representation. http://www.sei.cmu.edu/reports/02tr011.pdf. Accessed 11 Feb 2010

SEI (2002b) CMMI for systems engineering, software engineering, integrated product and process development, and supplier sourcing. Version 1.1—staged representation. http://www.sei.cmu.edu/reports/02tr012.pdf. Accessed 11 Feb 2010

SEI (2006a) Appraisal requirements for CMMI. Version 1.2. http://www.sei.cmu.edu/reports/06tr011.pdf. Accessed 16 Feb 2010

SEI (2006b) CMMI for development. Version 1.2. http://www.sei.cmu.edu/reports/06tr008.pdf. Accessed 3 Dec 2009

SEI (2006c) Standard CMMI appraisal method for process improvement (SCAMPI) A. Version 1.2—method definition document. http://www.sei.cmu.edu/reports/06hb002.pdf. Accessed 16 Feb 2010

SEI (2007) CMMI for acquisition. Version 1.2. http://www.sei.cmu.edu/reports/07tr017.pdf. Accessed 11 Feb 2010

SEI (2009) CMMI for services. Version 1.2. http://www.sei.cmu.edu/reports/09tr001.pdf. Accessed 11 Feb 2010

Sheard SA (1997) The frameworks quagmire. Crosstalk: J Def Softw Eng 10(9):9

Sheard SA (2001) Evolution of the frameworks quagmire. IEEE Comput 34(7):96–98

Shewhart WA (1980) Economic control of quality of manufactured product (Originally published in 1931), 2nd edn. American Society for Quality Control, Milwaukee

Shewhart WA (1986) Statistical method from the viewpoint of quality control (Originally published in 1939), 2nd edn. Dover Publications, New York

Simatupang TM, Sridharan R (2005) The collaboration index: a measure for supply chain collaboration. Int J Phys Distrib Logist Manag 35(1):44–62

Simon HA (1996) The sciences of the artificial, 3rd edn. Massachusetts Institute of Technology, Cambridge

Simon J-M, El Emam K, Rousseau S, Jacquet E, Babey F (1997) The reliability of ISO/IEC PDTR 15504 assessments. Softw Proc Improv Pract 3(3):177–188

Skjoett-Larsen T, Thernoe C, Andresen C (2003) Supply chain collaboration. Theoretical perspectives and empirical evidence. Int J Phys Distrib Logist Manag 33(6):531–549

Skrinjar R, Bosilj-Vuksic V, Stemberger MI (2008) The impact of business process orientation on financial and non-financial performance. Bus Proc Manag J 14(5):738–754

Skrinjar R, Stemberger MI, Hernaus T (2007) The impact of business process orientation on organizational performance. Proceedings of the 2007 informing science and IT education joint conference. Ljubljana, pp 171–185, 22–25 June 2007

Smith A, Mazlish B (2002) The wealth of nations. Representative selections (Originally published in 1776), 4rd edn. Dover Publications, New York

Smith H, Fingar P (2002, 2006) Business process management: the third wave. The breakthrough that defines competitive advantage for the next fifty years. Meghan-Kiffer Press, Tampa

Smith H, Fingar P (2004) Process management maturity models. http://www.bptrends.com/publicationfiles/07-04%20COL%20Maturity%20Models-%20Smith-Fingar.pdf. Accessed 23 June 2010

Spanyi A (2004a) Beyond process maturity to process competence. http://www.bptrends.com/publicationfiles/06-04%20ART%20Dev%20Business%20Process%20Competence%20-%20Spanyi.pdf. Accessed 23 June 2010

Spanyi A (2004b) Towards process competence. http://www.spanyi.com/images/BPM%20Towards.pdf. Accessed 6 Apr 2010

Spekman RE, Carraway R (2006) Making the transition to collaborative buyer-seller relationships: an emerging framework. Ind Mark Manage 35:10–19

SQI (2007) SPICE. http://www.sqi.gu.edu.au/spice/. Accessed 1 Mar 2010

Stevens GC (1989) Integrating the supply chain. Int J Phys Distrib Logist Manag 19(8):3–8

Supply-Chain Council (2008) Supply-chain operations reference-model. SCOR overview. Version 9.0. http://archive.supply-chain.org/galleries/public-gallery/SCOR%209.0%20Overview%20Booklet.pdf. Accessed 14 June 2010

Tan KC (2001) A framework of supply chain management literature. Europ J Purch Supply Manag 7:39–48

Tapia RS, Daneva M, van Eck P, Wieringa R (2008) Towards a business-IT alignment maturity model for collaborative networked organizations. Proceedings of the international workshop on enterprise interoperability, Munich, Germany, in conjunction with the 12th IEEE international EDOC conference, Munich, Germany. IEEE, Munich, pp 70–81, 15–19 Sept 2008

Thijs N, Bouckaert G (2007) Kwaliteit in de publieke sector: over het bos en de bomen. In: Thijs N, Bouckaert G (eds) Kwaliteit in beweging. Ervaringen met Kwaliteitsmanagement in Lokale Besturen. Vanden Broele, Brugge, pp 5–53

Tolsma J, de Wit D (2009) Effectief procesmanagement. Procesgericht sturen met het BPM model, 2nd edn. Eburon, Delft

Towers S (2009) 8 Omega ORCA. Organization readiness & competence assessment. http://www.slideshare.net/stowers/orca-overview. Accessed 20 June 2010

van den Bergh J (2010) Supply chain maturity scan. http://www.jvdbconsulting.com/supply-chain-maturity-scan.html. Accessed 23 June 2010

van Dooren W, Thijs N, Bouckaert G (2004) Quality management and the management of quality in European public administrations. In: Löffler E, Vintar M (eds) Improving the quality of East and West European public services. Ashgate Publishing, Hampshire, pp 91–106

Van Loon H (2004) Process assessment and ISO/IEC 15504. A reference book. Springer, New York van Steenbergen M, Bos R, Brinkkemper S, van de Weerd I, Bekkers W (2010) The design of focus area maturity models. Proceedings of DESRIST2010. Springer, St.Gallen, pp 317–332, 4–5 June 2010

VICS (2004) Collaborative planning, forecasting and replenishment (CPFR). An overview. http://www.vics.org/docs/standards/CPFR_Overview_US-A4.pdf. Accessed 14 June 2010

vom Brocke J, Rosemann M (2010) Foreword. In: vom Brocke J, Rosemann M (eds) Handbook on business process management, vol 2. Springer, Berlin Heidelberg, pp vii–ix

vom Brocke J, Sinnl T (2011) Culture in business process management: a literature review. Bus Proc Manag J 17(2):357–377

Walls JG, Widmeyer GR, El Sawy OA (2004) Assessing information system design theory in perspective: how useful was our 1992 initial rendition? JITTA: J Inf Technol Theory Appl 6(2):43–58

Weske M (2010) Business process management. Concepts, languages and architectures. Springer, Berlin Heidelberg

WfMC (1999) Terminology & glossary. http://www.wfmc.org/standards/docs/TC-1011_term_glossary_v3.pdf. Accessed 12 Jan 2011

Willaert P, Van den Bergh J, Willems J, Deschoolmeester D (2007) The process-oriented organisation: a holistic view. Developing a framework for business process orientation maturity. In: Alonso G, Dadam P, Rosemann M (eds) Business process management. 5th international conference, BPM 2007, Brisbane, Australia. Proceedings. LNCS 4714. Springer, Berlin Heidelberg, pp 1–15, 24–28 Sept 2007

Windle S (2004) Book review. Business process management (BPM): the third wave. J Inf Syst:128–130

Winter R (2008) Design science research in Europe. Europ J Inf Syst 17:470–475

Wognum PM, Faber EC (2002) Infrastructures for collaboration in virtual organisations. Int J Netw Virtual Organ 1(1):32–54

Yaibuathet K, Enkawa T, Suzuki S (2008) Influences of institutional environment toward the development of supply chain management. Int J Prod Econ 115:262–271

Yaibuathet K, Enkawa T, Yoshika T (2007) Impact of information technology and SCM organizational strategy on corporate financial performance and its cross national comparison. Int J Inf Syst Logist Manag 3(1):13–24

Zhao LJ, Cheng HK (2005) Web services and process management: a union of convenience or a new area of research? Decis Support Syst 40:1–8

zur Muehlen M, Ho DT-Y (2006) Risk management in the BPM lifecycle. In: Bussler CJ, Haller A (eds) BPM 2005 international workshops, BPI, BPD, ENEI, BPRM, WSCOBPM, BPS, Nancy, France, Proceedings. LNCS 3812. Springer, Berlin Heidelberg, pp 454–466, 5 Sept 2005

Chapter 2
Research

Abstract This chapter provides an executive summary of the three studies covered by this book, as an expansion of and redirection to journal articles (explicitly acknowledged by references). First, the literature study distinguished business process management (BPM) from business process orientation (BPO), in order to obtain a common understanding among researchers and practitioners. An essential difference in scope exists, with: (1) BPM being limited to the characteristics of business processes and the traditional business process lifecycle, and (2) BPO adding the organisation-specific characteristics to BPM, i.e. to make the organisational culture and structure more process-oriented. Furthermore, definitions were derived for maturity (levels), capability (levels) and for a business process maturity model (BPMM). Afterwards, the classification study identified the capability areas and maturity types of 69 existing BPMMs (see Chap. 1 for references), in order to strengthen the BPMM foundation. The maturity types were called: BPM maturity, intermediate BPO maturity (i.e. BPM plus culture), and BPO maturity (i.e. BPM plus culture and structure), and this for one, more or all business processes in an organisation. Finally, the selection study identified criteria for choosing one BPMM out of the wide array, resulting in a free and online decision tool (called BPMM Smart-Selector). Evaluation scores were calculated per collected BPMM, allowing an additional quality check.

Keywords Business process · Maturity · Capability · Business process management · Business process orientation · Continuous improvement · Excellence · Conceptual framework · Decision support system · Delphi technique · Analytical hierarchy process

2.1 Definitions

Reference to the literature study:

Van Looy A, De Backer M, Poels G (2011) Defining business process maturity. A journey towards excellence. Total Qual Manag Bus Excell 22(11):1119–1137. doi:10.1080/14783363.2011.624779

A. Van Looy, *Business Process Maturity*, SpringerBriefs in Business Process Management, DOI: 10.1007/978-3-319-04202-2_2, © The Author(s) 2014

2.1.1 Scope and Purpose

The introduction chapter explained that typically two maturity types are assumed to exist in the context of business processes: (1) the maturity of specific processes, and (2) the maturity of all processes in an organisation. The former is frequently called 'process maturity', whereas the latter is often referred to as 'business process management maturity' (de Bruin and Rosemann 2007). Nonetheless, at first sight, this traditional dichotomy was not clear when collecting existing BPMMs (see Chap. 1 for references to our sample with 69 BPMMs). It turned out that some BPMMs have model names that differ from the above, and that many BPMMs with the same model name partly measure different things. Also other concepts like 'maturity', 'capability', and 'maturity model' were used incoherently, and remained ambiguous as such. In order to solve these inconsistencies, the first study defined the terminology to be used in this book, in order to obtain a common understanding for researchers and practitioners.

- RQ1a. What is the appropriate BPMM scope? When does a maturity model consider business processes, and can be called a BPMM?
- RQ1b. What is the appropriate BPMM terminology? When does a maturity model consider maturity and capability?
- RQ1c. What is the appropriate BPMM design? When does a maturity model provide practical guidance to achieve process excellence?

2.1.2 Research Methodology

First, a literature study was conducted to define three umbrella terms in the process literature, to which BPMM names frequently refer: business process (BP), business process management (BPM) and business process orientation (BPO). For instance, examples of BPMM names are 'process maturity grid' (Harrington 2006), 'business process management maturity model' (de Bruin and Rosemann 2007), or 'business process orientation maturity model' (McCormack and Johnson 2001). Clear and recognised definitions were compared to find distinct components that clarify a difference in scope between these umbrella terms, and between BPMMs as such.

Afterwards, definitions were sought for 'maturity', 'maturity level', 'capability', and 'capability level' by primarily comparing the respective definitions in the main BPMM tracks (see Chap. 1), and by considering their meaning with regard to other literature on business process maturity.

A similar approach was taken to define a 'business process maturity model'. Besides definitions of the BPMM main tracks and literature on business process maturity, we relied on the general design elements of a maturity model.

2.1.3 Overview of the Findings

The literature study revealed that business process definitions implicitly focus on (1) **business process modelling** and (2) **deployment**. For instance, '*a process is a series of interconnected activities that takes input, adds value to it, and produces output. It's how organisations work their day-to-day routines. Your organisation's processes define how it operates*' (Harrington 2006: p. xxii). Deployment means performing or running processes in real-life, and can be emphasised by verbs like 'work' and 'operate'. It implicitly requires modelling or predefining business processes to clarify which inputs, activities and outputs are part of them. It concerns the first phases of a typical business process lifecycle (Weske 2010). The implicit assumptions of business process definitions are made explicit by BPM definitions.

BPM definitions additionally focus on (3) **business process optimisation** and (4) **management** by a process owner and a cross-functional process team. For instance, '*business process management includes concepts, methods, and techniques to support the* (1) *design,* (4) *administration,* (2) *configuration, enactment, and* (3) *analysis of business processes*' (Weske 2010: p. 5). This well-known and accepted definition clearly shows the four components, which conform to all phases of a typical business process lifecycle. They can be applied to one, more or all processes in an organisation.

Besides process characteristics, recent business process literature starts to recognise the importance of organisational characteristics (i.e. culture and structure) to obtain business (process) excellence (vom Brocke and Sinnl 2011). In order to stipulate this difference in scope, the notion of business process orientation (BPO) is introduced as an organisation that '*emphasises* (1–4) *process,* (5) *a process-oriented way of thinking, customers, and outcomes* (6) *as opposed to hierarchies*' (McCormack and Johnson 2001: p. 185). BPO theoretically combines the four BPM components with two BPO-specific components. The latter refer to the adoption of: (5) a **process-oriented culture**, e.g. with top management support and rewards linked to the performance of business processes instead of departments, and (6) a **process-oriented structure**, e.g. with a centre of excellence and a horizontal or matrix chart instead of vertical departments. These BPO-specific components impact the whole process portfolio of an organisation, whereas the BPM components remain limited to specific business processes (i.e. one, more or all).

A summary is visualised in Fig. 2.1 as a funnel structure with BPM being a subset of BPO. A maturity model considers business processes if it addresses these theoretical components, though not necessarily all these components. Depending on which of these components are actually addressed, a BPMM can deal with BPM maturity (i.e. modelling, deployment, optimisation or management, but not the organisational culture or structure) or BPO maturity (i.e. including the organisational culture or structure). Business process maturity as such (i.e. addressing merely modelling and deployment) is rather unlikely to exist, given the inherent

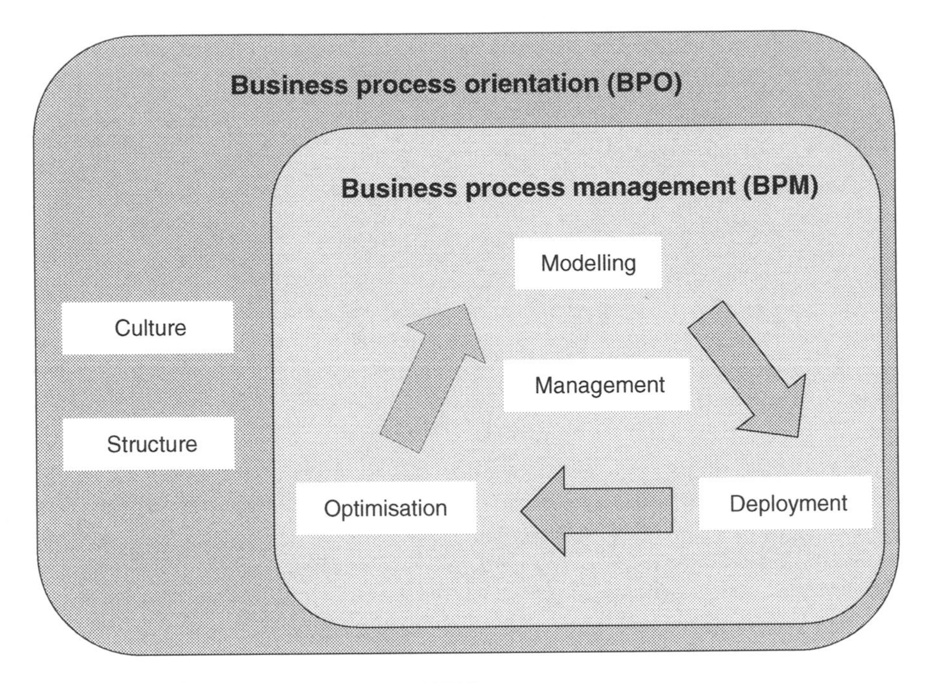

Fig. 2.1 The funnel structure of BPM and BPO

importance of continuous improvements (i.e. 'optimisation') to reach business (process) excellence.

Furthermore, the literature study clarified that these components are called capabilities (or rather areas of related capabilities) in the context of maturity models. They serve as a collection of necessary competences, skills or knowledge for an organisation to be capable of achieving the targeted process results. Maturity, on the other hand, considers the collection of capability areas in order to express '*the extent to which an organisation has explicitly and consistently deployed processes, according to their business objectives*' (Van Looy et al. 2011: p. 1129). This means that a capability level refers to the growth per capability area, whereas a maturity level refers to the overall growth of capability areas. The main BPMM tracks (see Chap. 1) often refer to 'organisational maturity' and 'process capability' to emphasise this difference in scope.

Consequently, we defined a business process maturity model as '*a model to assess and/or to guide best practice improvements in organisational maturity and process capability, expressed in lifecycle levels, by taking into account an evolutionary road map regarding* (1) *process modelling*, (2) *process deployment*, (3) *process optimisation*, (4) *process management*, (5) *the organisational culture, and/or* (6) *the organisational structure*' (Van Looy et al. 2011: pp. 1132–1133). This definition stipulates that, in order to be of practical use, a BPMM design should include both (1) an assessment method to determine actual levels, and (2) an improvement method with a road map to reach desired levels.

2.1.4 Recommendations

This study defined the fundamentals of a business process maturity model (BPMM). We recommend scholars and practitioners to apply the defined concepts accordingly, in order to obtain a common understanding and clear communication. For instance, the sampled BPMMs were created without much BPMM-specific literature to guide them, which explains some inconsistencies among them. It also explains why BPMMs are expected to differ in quality. One of these quality differences concerns the degree to which BPMMs provide guidance. Hence, we recommend the use of a BPMM with both a detailed assessment method and improvement method. Study 3 continues to elaborate on this BPMM choice. Additionally, we recommend practitioners to strive for an optimal, desired maturity level or capability levels, depending on their organisational needs and context (instead of blindly striving for the highest levels).

Of paramount importance is our unique arrangement of BPM and BPO components into a funnel structure, suggesting different types of maturity. Particularly, the current business process literature refers to BPM and BPO as synonyms, without stipulating a difference in scope. Instead, we refined those two umbrella terms by explicitly recognising their distinct components and to relate them accordingly. We recommend researchers to adopt the same critical reflection, for the scope of their research being clear at a glance.

Acknowledgments We acknowledge the written permissions of ISO/IEC, OMG, and SEI to cite their definitions for academic purposes. We also note that Capability Maturity Model and CMM are registered trademarks in the U.S. Patent and Trademark Office. CMM Integration and CMMI are service marks of Carnegie Mellon University.

2.2 Classification

Reference to the classification study:

Van Looy A, De Backer M, Poels G (2012) A conceptual framework and classification of capability areas for business process maturity. Enterp Inf Syst. doi:10.1080/17517575.2012.688222

2.2.1 Scope and Purpose

The classification study elaborated on the distinct components of BPM and BPO (study 1) by decomposing them into sub components. This specification was required to enable a rigorous mapping of the capability areas in existing BPMMs to their theoretical equivalents, and to find evidence of a process capability framework. We recall that maturity refers to the expected performance, which is an indicator of the actual performance. Consequently, process capability areas are assets or critical success factors for business (process) performance, and can be underpinned in the broader literature. However, no consensus exists on the process capability areas. Many scholars have (mostly empirically) examined the critical success factors for business (process) excellence, which frequently resulted in a new BPMM (Hammer 2007; Harrington 1991; McCormack and Johnson 2001; de Bruin and Rosemann 2007). Unlike a sound methodology or use, these alternative solutions are not grounded by established theories and still differ in the capability areas covered. Therefore, this study looked for theories that underlie the process capability areas in order to create a BPMM-independent foundation. The intended conceptual framework not only serves our research on business process maturity, but helps to consolidate and advance the business process literature.

The ultimate aim of this study was to verify whether different maturity types are being measured by existing BPMMs, both by referring to what they assess and improve (i.e. the capability areas, study 1) and to their number of business processes addressed (i.e. as suggested by de Bruin and Rosemann (2007)). Consequently, this study intended to add a novel perspective to the literature of business process maturity by verifying the traditional dichotomy of maturity types and by supplementing this division with well-grounded capability areas.

- RQ2a. Which capability areas can be assessed and improved by a BPMM in order to reach business (process) excellence?
- RQ2b. Which capability areas are actually assessed and improved by existing BPMMs?
- RQ2c. If RQ2b shows that different capability areas are actually assessed and improved, do existing BPMMs measure different types of maturity?

2.2.2 Research Methodology

As a continuation of study 1, a literature study was conducted to theoretically identify the business process capability sub areas. Houy et al. (2010) argue that the business process literature is mainly empirically based, with currently a poor tradition of theory construction. To our knowledge, only the business process lifecycle theories are well established, which are mostly limited to 'modelling', 'deployment', and 'optimisation'. Therefore, we concretised the other main capability areas by relying on the broader business process literature, among others the business process evolutions of Chap. 1 and review articles that differ from maturity models (Kohlbacher 2010; Lee and Dale 1998; Palmberg 2009). We underpinned these findings by organisation management theories regarding: (1) performance and change management, (2) human resource management, and (3) strategic management. Further on, the resulting main areas and sub areas were empirically validated by mapping our sample of 69 BPMMs (see Chap. 1) to them.

Regarding the maturity types, a classification study was conducted by combining exploratory cluster analysis (i.e. to identify maturity types) with confirmatory discriminant analysis (i.e. to validate maturity types) (Punj and Stewart 1983; Romesburg 1984). As an exploratory classification technique, any cluster analysis tends to produce a classification, regardless of whether data actually comprise natural groupings (Punj and Stewart 1983). This requires some precautions to avoid a solution that just occurred '*by chance or as an artefact of a clustering algorithm*' (Jain et al. 1999: p. 268). For this purpose, Romesburg (1984) advises to choose the most meaningful clustering solution, which generates interesting and useful conclusions, and which is stable on both the complete and split dataset. All clustering methods in SPSS (version 18) were evaluated in this respect. A Cohen's Kappa was computed as a measure of agreement on group membership between solutions, e.g. between complete and split datasets or between cluster and discriminant analysis.

Cluster analysis can be conducted without assumptions about the underlying data distribution. On the other hand, discriminant analysis is enhanced by seven assumptions (Klecka 1980), which were translated to our research as follows: (1) at least two clusters, (2) at least two BPMMs per cluster, (3) the number of discriminating variables (i.e. capability sub areas) is less than the total number of BPMMs minus 2 (i.e. N = 69–2), (4) the discriminating variables are at the interval level or binary, (5) non-linear relationships between discriminating variables, (6) homogeneity of variance, i.e. equal variance within each of the clusters, and (7) a multivariate normal distribution on the discriminating variables. Our discriminating variables are expected to be binary (i.e. sub areas are present or absent in the BPMM design documents), and thus likely to dissatisfy the sixth and seventh assumption. Nonetheless, Klecka (1980) asserts that a discriminant analysis can still be performed for binary data. Only in worst case, when the discriminant functions are not statistically significant, the intended discriminant analysis cannot be used as a classification technique.

2.2.3 Overview of the Findings

Figure 2.2 illustrates the resulting conceptual framework. On the left, it shows the sub areas per main area of study 1. For a description per sub area, we refer to Van Looy et al. (2012).

The six sub areas related to 'modelling', 'deployment', and 'optimisation' were directly derived from the traditional business process lifecycle theories. Although many lifecycle variants exist, they do not fundamentally differ (Shewhart 1986; Deming 1994; Harrington 1991; Kannengiesser 2008; Netjes et al. 2006; Smith and Fingar 2002, 2006; Weske 2010; zur Muehlen and Ho 2006). The main difference is that some variants also mention the 'management' of business processes, albeit without considering all aspects (Weske 2010; zur Muehlen and Ho 2006). This is where organisation management theories come to the foreground.

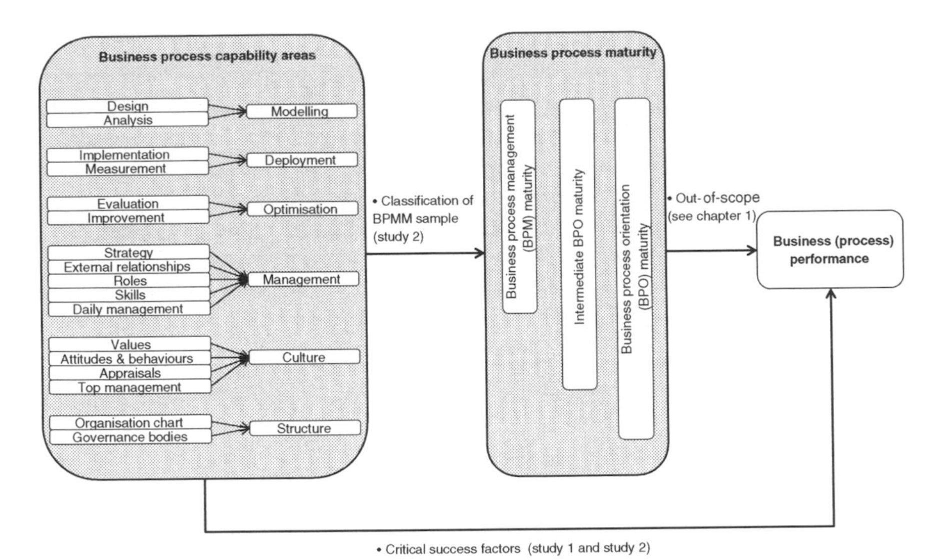

Fig. 2.2 The process capability framework and types of business process maturity

First, the organisation performance and change theories of Waterman et al. (1980) and Burke and Litwin (1992) (causally) describe aspects that affect business performance. Although we acknowledge differences between business performance and process performance, all theoretical aspects affect our six main areas and were mapped to our 17 sub areas.

Secondly, theories on strategic management emphasise that the organisational strategy must be translated into the strategy of a specific business process (i.e. 'management'). For instance, a balanced scorecard systematically derives key performance indicators regarding four perspectives of business performance, with business processes being one of the perspectives (Kaplan and Norton 2001).

Thirdly, the human resource management theory of Boswell et al. (2006; Colvin and Boswell 2007) impacts both the 'management' and 'culture' areas. Particularly, this theory argues that the behaviours of employees can be directed: (1) by obtaining the skills and knowledge to perform a single process (i.e. 'management'—'skills') and (2) by obtaining the motivation to perform through top management communication, employee involvement in decision-making, extrinsic motivation and intrinsic motivation (i.e. 'culture'—'top management', 'attitudes and behaviours', 'appraisals', 'values' respectively).

Content analysis validated the comprehensiveness of the conceptual framework by mapping all capability areas of 69 BPMMs to a theoretical equivalent. No other sub areas were present in the sample. Appendix illustrates how a detailed mapping was established per sampled BPMM. This mapping showed many similarities with alternative process capability frameworks. For instance, an almost perfect comparison could be made with the process capability areas of de Bruin and Rosemann (2007), except for the organisation chart. Their BPMM was thoroughly designed by using international Delphi studies, and validated by case studies. It consists of six main areas, each with five sub areas: (1) strategic alignment, (2) governance, (3) methods, (4) information technology, (5) people, and (6) culture. Our conceptual framework consolidates their findings with other BPMMs, and, most importantly, complements them with underlying theories.

Descriptive statistics showed, however, that BPMMs usually do not cover all theoretical areas, which makes classification worthwhile to detect maturity types. Only the Ward's and k-means methods with three clusters satisfied the classification assumptions, mentioned in the methodology section. Since both solutions statistically fit our data (on both the complete and different split datasets), the final clustering was guided by its meaningfulness. Ward divides the 69 BPMMs in three clearly separated clusters of almost equal size (i.e. 23, 20, and 26 BPMMs): a partial BPM cluster (i.e. with some modelling, deployment, optimisation, and management areas), a quasi-full BPO cluster (i.e. with almost all areas), and an intermediate cluster (i.e. by combining BPM with some BPO-specific capability areas). On the other hand, k-means proposes unequal clusters (i.e. 15, 16 and 38 BPMMs) with a more ambiguous cluster membership: a minimal BPM cluster (i.e. with some modelling, deployment and optimisation areas), a partial BPM cluster (i.e. by also including some management areas), and a partial BPO cluster (i.e. by also including culture-related areas and some structure-related areas). As the k-means method has a less distinct representation of the capability areas per cluster, the Ward's method was preferred for our study. The adequacy of the three resulting clusters was judged by the discriminant methods available in SPSS (version 18), i.e. the regular and stepwise methods. As the vast majority of the sampled BPMMs were predicted in the same clusters as in cluster analysis, the BPMM classification was strongly confirmed. Consequently, we obtained a reliable classification solution, which appeared to guarantee cross-validation by adhering to our prior expectations of BPM maturity and BPO maturity (study 1) (Punj and Stewart 1983; Romesburg 1984). It also refined the funnel structure of our previous study by revealing a third and yet unknown intermediate cluster.

In sum, the three maturity types resulting from the classification study are:

- **business process management (BPM) maturity**, primarily focusing on business process modelling (1), deployment (2), optimisation (3) and management (4);
- **business process orientation (BPO) maturity**, combining BPM maturity with a process-oriented culture (5) and structure (6);
- **intermediate BPO maturity**, limiting BPO maturity to some process-oriented aspects, usually cultural (5).

These maturity types are, however, strictly limited to the capability coverage of sampled BPMMs. They do not take into account the number of business processes addressed. This variable was, however, also registered during content analysis. We noticed that the assessment items in BPMMs literally refer to one, more or all business processes in an organisation, and this for BPMMs of all maturity types. Consequently, nine maturity types appeared to exist:

- **BPM** maturity for **one, more or all** business processes;
- **Intermediate BPO** maturity for **one, more or all** business processes;
- **BPO** maturity for **one, more or all** business processes;

The advanced maturity types refine the dichotomy of de Bruin and Rosemann (2007) by dividing specific processes in either a single process or either a business domain with multiple (sub) processes (e.g. a value chain). We also argued that these findings allow to compare the completeness of BPMMs. Such a critical view on BPMMs turned out to be crucial for the sake of clarity, as their names not always correctly represent the exact coverage (i.e. with possible over- or under-estimates). For instance, some BPMM names refer to 'BPM maturity', whereas they actually measure BPO maturity for all processes (i.e. being more complete), or vice versa, etc. Nevertheless, the best suited maturity type for a particular organisation is not necessarily the most complete one. Future research is required to investigate which contextual factors determine the choice of maturity type, e.g. based on the degree of top management support, etc.

2.2.4 Recommendations

This study theoretically and empirically strengthened the findings of study 1. It responded to the lacking consensus on capability areas necessary for business process excellence, and grounded the BPMM literature to some degree. Internal validity and reliability were obtained by an iterative and top-down approach for identifying and validating both capability areas and maturity types. As our BPMM sample includes different process types (i.e. generic, supply chains, collaboration), it suggests versatility and external validity to some extent. External validity is ensured when recognising the classification utility in a larger sense, i.e. as a first and essential step towards a theory of business process maturity. Researchers are encouraged to further examine the effects of capability areas and maturity types on business (process) performance. Hence, our conceptual framework has the potential to contribute to a theoretical framework in future research (see Fig. 2.2 and Chap. 3).

Study 2 illustrated that blindly comparing results from different BPMMs (e.g. for benchmarking) is not justified, as not all BPMMs measure the same maturity type. We recommend scholars and practitioners to adopt a critical view on the BPMM names by applying our nine business process maturity types. These maturity types are also more informative than the traditional dichotomy of maturity types, and therefore more recommended. Furthermore, our maturity types are helpful to practitioners in order to interpret the direction of BPMMs, i.e. which capability areas are to be assessed and improved, and to motivate their choice for one or another maturity type based on their organisational needs. We strongly recommend practitioners to choose a maturity type that best fits their organisational context, instead of blindly choosing one maturity type or ambitiously taking the most complete one. Finally, researchers are encouraged to investigate these contextual factors in more detail.

2.3 Selection

Reference to the selection study:

Van Looy A, De Backer M, Poels G, Snoeck M (2013) Choosing the right business process maturity model. Inf Manag. doi:10.1016/j.im.2013.06.002

2.3.1 Scope and Purpose

The third study continued to focus on the great BPMM variety by advising on which BPMM must be chosen when, based on discussions with international subject-matter experts. This study broadened our BPMM comparison from capability areas and number of business processes to other BPMM characteristics. It evaluated the sampled BPMMs with regard to more aspects than merely our BPMM classification (by means of evaluation scores), and particular organisations can use our findings to evaluate which BPMMs best fit their needs (by means of an online decision tool). The emphasis is on BPMM selection as a first essential phase before BPMM application, given the multitude of diverse BPMMs.

- RQ3a. Which criteria help users (i.e. organisations or academics) choose a BPMM?

2.3.2 Research Methodology

The design-science paradigm was followed to develop and test a BPMM decision tool, called BPMM Smart-Selector. A solution was proposed by relying on: (1) the IS design research cycle and guidelines of Hevner et al. (2004), (2) the IS artefact types of March and Smith (1995), and (3) the IS design theory components of Walls et al. (2004). The design hypotheses (or requirements) were formulated based on the following decision-making theories: (1) the theory of bounded rationality (Simon 1979), (2) the theory of information symmetry (Afzal et al. 2009), (3) the theory of managerial work (Mintzberg 1971), and (4) the multi-attribute utility theory (Keeney and Raiffa 1993). The tool operates like a decision table, and is freely available at: http://smart-selector.amyvanlooy.eu/.

The selection criteria to be included in the BPMM decision tool were discussed and chosen by means of an international Delphi study (i.e. consensus-seeking decision-making) (Dalkey and Helmer 1963), and weighed by the Analytical Hierarchy Process (AHP, i.e. multi-attribute decision-making) (Saaty 1990). Both methods are widely used for exploring ideas and ratings.

More specifically, we applied a 'modified' Delphi approach, which started from an initial list of criteria based on content analysis (see Chap. 1), peer feedback and a pilot study. Experts could give open comments at any time and propose an unlimited list of missing criteria. As this approach typically gives common ground, it allowed including experts of different backgrounds to stimulate normative discussions. In total, 11 BPM academics and 11 BPM practitioners participated in the study, each from five continents. They were carefully chosen conform to Okoli and Pawlowski (2004), and remained anonymous during the study. Only two experts permanently dropped-out after round 1 due to other obligations.

The duration of the Delphi study was directed by strict conditions regarding consensus, stability and fatigue. It took five rounds between November 2011 and April 2012, i.e. one round for brainstorming possible criteria, two for narrowing down, one for weighing the final criteria by AHP, and a wrap-up round.

2.3.3 Overview of the Findings

In total, 24 criteria and their trade-offs were discussed by the subject-matter experts (Table 2.1). As these criteria concern BPMM characteristics, they were reorganised into the conceptual model of a BPMM to refine study 1. For a description per criterion and their trade-offs, we refer to Van Looy et al. (2013).

On the left, Table 2.1 presents the 14 criteria that reached consensus to be included in the BPMM Smart-Selector. These criteria and their options were weight by using AHP to know their relative importance, i.e. according to the expert panel. In line with study 1 and study 2, the capability coverage of BPMMs received the highest weight, which means that the experts confirmed the criterion's utmost importance. Moreover, successive warm-up discussions showed that the statistical clusters of study 2 (i.e. maturity types) were experienced quite naturally by almost the entire expert panel. In order to give more guidance on the choice of capability cluster (study 2), two additional contextual factors were derived from the expert discussions: (1) the degree of IT support, and (2) BPM experience.

On the right, Table 2.1 presents the ten criteria that did not reach consensus, and were omitted from the questionnaire in the BPMM Smart-Selector. These criteria did not show a bimodal distribution with academics opposed to practitioners, but were blocked by a small minority with opposite opinions.

Subsequently, two evaluation scores were calculated per sampled BPMM, as an additional quality check before launching the tool: (1) a selection score, and (2) a transparency score. The selection score considered the 14 decision criteria that reached consensus, and assigned the weighed options to BPMMs based on content analysis (Chap. 1). To avoid bias, the mapping of BPMMs to criteria was conducted before the weights were defined, and not known by the experts when weighing. On the other hand, a transparency score verifies whether all 24 criteria considered in the Delphi study (i.e. without consensus to be excluded from the tool) are present in the BPMM design documents. In other words, once a BPMM selected, does the user get sufficient information to start using it? Based on these evaluation scores, nine BPMMs were omitted from the BPMM Smart-Selector.

Table 2.1 The criteria discussed in the Delphi study

Criteria with consensus to be included in the tool (in the order of importance)	Criteria without consensus to be in- or excluded
• Presence of capabilities	• Calculation of (maturity or capability) levels
• Architecture type	• Representation of (maturity or capability) levels
• Architecture details	• Number of assessed organisations
• Type of business processes	• Lead assessor
• Rating scale	• Number of assessors
• Data collection technique	• Business versus IT
• Purpose	• Number of (maturity or capability) levels
• Validation	• Labelling of levels
• Number of business processes	• External levels
• Assessment duration	• Methodology
• Assessment availability	
• Functional role of respondents	
• Number of assessment items	
• Direct costs	

Consequently, the prototype of the BPMM Smart-Selector (http://smart-selector.amyvanlooy.eu/) consists of a questionnaire with 14 selection criteria which guide the user through 60 (instead of 69) sampled BPMMs. The questionnaire presents the criteria (i.e. questions with trade-offs) in the order of importance according to our expert panel. The user, however, can also start with those questions that are most relevant to his organisation (and use other questions to refine the results afterwards) instead of following the proposed sequence. An initial version was tested by employees enrolled for the BPM course of a post-graduate training program (as potential users). Afterwards, the prototype tool was successfully applied in three case studies: (1) a profit organisation not yet using a BPMM, (2) a non-profit organisation already using a BPMM, and (3) by an academic for research purposes.

2.3.4 Recommendations

The critical view on BPMMs was continued in this third study. In contrast to the previous studies in this book, it directly resulted in practical advice by means of the BPMM Smart-Selector. Although Delphi studies are typically limited by their expert choice and design, several precautions were taken to ensure a sound methodology. Internal validity resulted from a careful selection of 22 experts (by role and region) and four coders (including one coder from another university to avoid bias) who participated throughout successive rounds. The resulting selection criteria turned out to be reliable, with high satisfaction rates among both the experts and the independent testers afterwards. Regarding external validity, the methodology ensured that the BPMM Smart-Selector can be updated with new

BPMMs or new criteria. Reuse for other types of decision tools (e.g. other maturity models) needs to be investigated.

We recommend practitioners to use the BPMM Smart-Selector in order to discover which BPMM best fits their organisational needs (http://smart-selector.amyvanlooy.eu/). The tool presents detailed information per criterion, including trade-offs. Organisations that wish to start with maturity assessments will be informed about their best matching BPMM in our sample, including references on how to access the model. On the other hand, organisations that already use a BPMM can evaluate whether that model would also be their best matching model, and possibly change over. Also scholars are recommended to use the BPMM Smart-Selector, for instance, to get to know them or if they consider including BPMMs in their research.

We must, however, note that the BPMM Smart-Selector concerns a 2012 proof-of-concept to validate its way or working. It is based on BPMM design documents of 2010 or earlier. The author of this book can be contacted if you wish to include additional BPMMs or corrections to BPMM details in a next release. Furthermore, we invite practitioners and researchers to give us feedback after using the BPMM Smart-Selector. To enter the feedback form, a button 'Ready? Give us your feedback' is present in the selection table (i.e. above the overview of matching BPMMs). The BPMM Smart-Selector also requests some anonymous data about you and your organisation to track how the tool is used. We guarantee that this data is merely used for academic purposes. For instance, such data collection can be used to explain which organisation types choose for a specific capability cluster (see study 2).

Acknowledgments We truly thank the coders and the expert panel for their continuing participation throughout the different Delphi rounds. Furthermore, we thank the testers for using our prototype tool.

Appendix: An Illustration of the Detailed BPMM Mapping

Please note that a detailed mapping per BPMM can be found at http://www.amyvanlooy.eu/research/2-classification

ID	BP modelling	BP deployment	BPM optimisation	BPM management	BPO culture	BPO structure	Other
OMG	Work unit requirements management (1/2 items) Organisational process management (1/3 items) Product and service preparation (3 items)	Product and service deployment (3 items) Product and service operations (3 items) Product and service support (3 items) Product and service process integration (1/3 items) Organisational improvement deployment (1/3 items) Work unit monitoring and control (3 items)	Work unit performance (3 items) Organisational process management (2/3 items) Product and service work management (2/3 items) Organisational capability and performance management (3 items) Product and service process integration (1/3 items) Quantitative product and service management (2 items) Quantitative process management (3 items) Defect and problem prevention (3 items) Continuous capability improvement (2 items) Organisational innovative improvement (3 items) Organisational improvement deployment (2/3 items)	Work unit requirements management (1/2 items) Work unit planning and commitment (3 items) Work unit configuration management (3 items) Organisational configuration management (3 items) Organisational resource management (2 items) Sourcing management (3 items) Process and product assurance (2 items) Organisational competency development (2 items) Product and service business management (3 items) Product and service work management (1/3 items)	Organisational business governance (2 items) Organisational process leadership (2 items) Organisational common asset management (2 items) Organisational performance alignment (2 items) Organisational improvement planning (3 items) Organisational process leadership (1 items)	Product and service process integration (1/3 items)	–
	Design: OK **Analysis: OK**	**Implementation: OK** **Measurement: OK**	**Evaluation: OK** **Improvement: OK**	**Strategy: OK** **External: OK** **Roles: OK** **Training: OK** **Daily management: OK**	**Values:-** **Attitudes: OK** **Rewards: OK** **Top management: OK**	**Chart: OK** **Bodies: OK**	

References

Afzal W, Roland D, Al-Squri MN (2009) Information asymmetry and product valuation: an exploratory study. J Inf Sci 35(2):192–203

Boswell WR, Bingham JB, Colvin AJ (2006) Aligning employees through 'line of sight'. Bus Horiz 49:499–509

Burke WW, Litwin GH (1992) A causal model of organizational performance and change. J Manag 18(3):523–545

Colvin AJ, Boswell WR (2007) The problem of action and interest alignment: beyond job requirements and incentive compensation. Hum Resour Manag Rev 17:38–51

Dalkey N, Helmer O (1963) An experimental application of the Delphi method to the use of experts. Manage Sci 9(3):458–467

de Bruin T, Rosemann M (2007) Using the Delphi technique to identify BPM capability areas. In: Proceedings of the 18th Australasian conference on information systems, Toowoomba, pp 642–653, 5–7 Dec 2007

Deming WE (1994) The new economics: for industry, government, education, 2nd edn. Center for Advanced Educational Services, Cambridge

Hammer M (2007) The process audit. Harv Bus Rev (4):111–123

Harrington HJ (1991) Business process improvement. The breakthrough strategy for total quality, productivity, and competitiveness. McGraw-Hill, New York

Harrington HJ (2006) Process management excellence. The art of excelling in process management. Book 1 in the five-part series. The five pillars of organizational excellence. Paton Press, California

Hevner AR, March ST, Park J, Ram S (2004) Design science in information systems research. MIS Q 28(1):75–105

Houy C, Fettke P, Loos P (2010) Empirical research in business process management—analysis of an emerging field of research. Bus Proc Manag J 16(4):619–661

Jain AK, Murty MN, Flynn PJ (1999) Data clustering: a review. ACM Comput Surv 31(3):264–323

Kannengiesser U (2008) Subsuming the BPM life cycle in an ontological framework of designing. In: Proceedings of the CIAO! and EOMAS workshops, CAiSE. Springer, Berlin Heidelberg, pp 31–45

Kaplan RS, Norton DP (2001) The strategy-focused organization. How balanced scorecard companies thrive in the new business environment. Harvard Business School Press, Boston

Keeney RL, Raiffa H (1993) Decisions with multiple objectives. Preferences and value tradeoffs. Cambridge University Press, Cambridge

Klecka WR (1980) Discriminant analysis. Sage Publications, California

Kohlbacher M (2010) The effects of process orientation: a literature review. Bus Proc Manag J 16(1):135–152

Lee RG, Dale BG (1998) Business process management: a review and evaluation. Bus Proc Manag J 4(3):214–225

March ST, Smith GF (1995) Design and natural science research on information technology. Decis Support Syst 15(4):251–266

McCormack K, Johnson WC (2001) Business process orientation: gaining the e-business competitive advantage. St. Lucie Press, Florida

Mintzberg H (1971) Managerial work: analysis from observation. Manage Sci 18(2):B97–B110

Netjes M, Reijers HA, van der Aalst WM (2006) Supporting the BPM life-cycle with FileNet. In: Proceedings of the EMMSAD workshop, CAiSE, Luxembourg. Namur University Press, Luxembourg, pp 497–508, 5–6 June 2006

Okoli C, Pawlowski SD (2004) The Delphi method as a research tool: an example, design constructions and applications. Inf Manag 42:15–29

Palmberg K (2009) Exploring process management: are there any widespread models and definitions? TQM J 21(2):203–215

Punj G, Stewart DW (1983) Cluster analysis in marketing research: review and suggestions for application. J Mark Res 20(2):134–148

Romesburg CH (1984) Cluster analysis for researchers. Lulu Press, North Carolina

Saaty TL (1990) An exposition of the AHP in reply to the paper remarks on the analytical hierarchy process. Manag Sci 36(3):259–268

Shewhart WA (1986) Statistical method from the viewpoint of quality control (originally published in 1939, 2nd edn. Dover Publications, New York

Simon HA (1979) Rational decision making in business organizations. Am Econ Rev 69(4):493–513

Smith H, Fingar P (2002, 2006) Business process management: the third wave. The breakthrough that defines competitive advantage for the next fifty years. Meghan-Kiffer Press, Tampa

Van Looy A, De Backer M, Poels G (2011) Defining business process maturity. A journey towards excellence. Total Qual Manag Bus Excell 22(11):1119–1137. doi:10.1080/14783363.2011.624779

Van Looy A, De Backer M, Poels G (2012) A conceptual framework and classification of capability areas for business process maturity. Enterp Inf Syst. doi:10.1080/17517575.2012.688222

Van Looy A, De Backer M, Poels G, Snoeck M (2013) Choosing the right business process maturity model. Inf Manag. doi:10.1016/j.im.2013.06.002

vom Brocke J, Sinnl T (2011) Culture in business process management: a literature review. Bus Proc Manag J 17(2):357–377

Walls JG, Widmeyer GR, El Sawy OA (2004) Assessing information system design theory in perspective: how useful was our 1992 initial rendition? JITTA: J Inf Techn Theory Appl 6(2):43–58

Waterman JR, Peters TJ, Phillips JR (1980) Structure is not organization. Bus Horiz (June):14–26

Weske M (2010) Business process management. Concepts, languages and architectures. Springer, Berlin Heidelberg

zur Muehlen M, Ho DT-Y (2006) Risk management in the BPM lifecycle. In Bussler CJ, Haller A (eds) BPM 2005 International workshops, BPI, BPD, ENEI, BPRM, WSCOBPM, BPS, Nancy, France. Proceedings. LNCS 3812. Springer, Berlin Heidelberg, pp 454–466, 5 Sept 2005

Chapter 3
Conclusion

Abstract Readers who prefer a high-level discussion of the results presented in this book, must certainly read this chapter. It recaps a comprehensive guide for organising and classifying the business process management discipline by means of a process capability framework. The latter gives specific attention to the organisational characteristics (culture and structure) that complement the process lifecycle, and so differs 'business process management' (BPM) from 'business process orientation' (BPO). The different maturity types resulting from the conceptual framework teach the readers to adopt a critical attitude towards business process maturity models (BPMMs). In order to solve the problem that many BPMMs exist, this chapter recalls the tips and tricks on how to choose the right maturity model for your needs. The BPMM Smart-Selector (http://smart-selector.amyvanlooy.eu/) acts as a free and online step-by-step tutorial (including trade-offs) to help the readers make a fast and reliable selection out of our unique and large BPMM sample (see Chap. 1 for references). Additionally, the scientific and managerial implications of the book are specified in detail. The chapter concludes by reflecting on the research limitations, and on how they can be addressed in future research. A call to researchers is made for collaborating and transforming the conceptual framework of this book into a theoretical framework that advances the business process management discipline.

Keywords Business process · Maturity · Capability · Results · Discussion · Contribution · Implications · Limitations · Future research

This final chapter concludes by discussing the findings of the book in Sect. 3.1. We hereby respond to our main research question stipulated in the introduction chapter. Section 3.2 specifies our scientific contributions, whereas the practical implications of our research are presented in Sect. 3.3. We also faced some research limitations, which are explained in Sect. 3.4. Finally, we elaborate on the next steps towards a theory of business process maturity and give some suggestions for other interesting research avenues in Sect. 3.5.

A. Van Looy, *Business Process Maturity*, SpringerBriefs in Business Process Management, DOI: 10.1007/978-3-319-04202-2_3, © The Author(s) 2014

3.1 Discussion of the Results

In this book, we have gained insight into the concept of business process maturity. The different studies constitute a comparative study based on a large sample of existing BPMMs (N = 69). To create a solid BPMM foundation, the studies were underpinned by literature on maturity models, business processes and organisation management.

Due to the lack of a common BPMM understanding, study 1 started with defining the BPMM scope (i.e. six components of business process maturity to be addressed), the BPMM terminology (i.e. the difference between maturity and capability) and the appropriate BPMM design (i.e. as an assessment and improvement method). The BPMM scope was derived from definitions for a business process, BPM, and BPO. The BPMM terminology and design were mainly derived by comparing the definitions for four main tracks of BPMMs in the BPMM literature, i.e. CMMI, FAA-iCMM, ISO/IEC 15504, and OMG-BPMM. These main tracks only focus on the maturity of specific processes, instead of also coping with all business processes in the organisation. Nevertheless, the small amount of existing literature on maturity for all business processes was included to generalise the findings. We demonstrated that business process maturity is rather an abstract concept, representing a set of capabilities that are needed for a business process to perform excellently. Maturity levels indicate the overall growth through all capabilities, whereas capability levels indicate the growth per capability. In order to provide practical guidance on business process excellence, a BPMM both assesses (AS-IS) and improves (TO-BE) business process maturity, instead of being limited to one of these crucial functions. The findings of study 1 were bundled in a BPMM definition, which was used throughout the research. Furthermore, it turned out that the six identified components act as capabilities, or rather capability areas for business processes (i.e. collections of related capabilities). They were adopted in study 2 as main capability areas, and subdivided into 17 capability sub areas.

In order to compare our BPMM sample with regard to the business process capability areas, study 2 first derived a conceptual framework from the literature and theories of the traditional business process lifecycle, performance and change management, human resource management, and strategic management. The capability areas (or components of business process maturity) of study 1 are central to this conceptual framework, particularly business process (1) modelling, (2) deployment, (3) optimisation, (4) management, and a process-oriented (5) culture, and (6) structure. Statistical classification techniques (i.e. exploratory cluster analysis and confirmatory discriminant analysis) revealed that BPMMs do not always address all capability areas, but that three clusters exist: (1) BPMMs limited to the first four areas (which we refer to as 'basic' capability areas for business processes), characterising the traditional business process lifecycle, (2) BPMMs combining the basic areas with a process-oriented culture, and (3) BPMMs addressing the basic areas plus a process-oriented culture and structure. This BPMM classification was confirmed in study 3, in which BPM experts experienced the statistical clusters quite naturally. Advice on which cluster to choose was limited to top management support in study 2, and broadened towards IT support and BPM experience in study 3. Future research is required to elaborate on such contextual factors.

Further on, de Bruin and Rosemann (2007) distinguish two types of business process maturity, which we refined in study 2: the maturity of specific business processes, and the maturity of BPM in general (i.e. the mastery of managing all business processes in the organisation). When analysing our sample, we noticed that the assessment items (i.e. questions to assess or measure capability areas) in BPMMs for specific business processes either literally refer to a single business process or to a particular business domain, such as a value chain with multiple (sub) processes. Hence, based on the terminology used in the assessment items of existing BPMMs, we proposed a distinction between (1) one business process, (2) more than one, but not all business processes in the organisation, and (3) all business processes in the organisation. BPMMs in the first situation are less numerous, but they give assessors more freedom to define the process boundaries themselves, e.g. is a business process assessed and improved as a sub process or as a separate process. The refinement of maturity types in study 2 not only copes with the number of business processes, but also adds our three statistical clusters based on the capability coverage. This boils down to nine different types of business process maturity, measured by the sampled BPMMs: BPM maturity for one, more or all processes (i.e. cluster 1), intermediate BPO maturity for one, more or all processes (i.e. cluster 2), and BPO maturity for one, more or all processes (i.e. cluster 3). Future research must investigate the pros and cons of using different subsets of business processes and different clusters of capability areas, and how they relate to each other.

In study 3, the comparative study of our BPMM sample was broadened to other BPMM characteristics. The purpose of this final study was to derive criteria that help users choose a BPMM out of the wide array, and to summarise the findings into a BPMM decision tool (called BPMM Smart-Selector). We therefore conducted an international Delphi study with 11 academics and 11 practitioners, each from five continents. Of the 24 decision criteria considered by the Delphi experts, 14 criteria reached consensus to be included in the tool and were weighed. The resulting criteria were translated into a questionnaire, starting with criteria of higher weights. By filling out the questionnaire, users will be navigated to the BPMMs that best fit their responses. This navigation is guided by a decision table, incorporated in the BPMM Smart-Selector, to process the responses. The decision table typically finds the best matching BPMM based on the sequence of answered questions (i.e. either the sequence proposed by the weights of the Delphi experts or a sequence determined by the user's particular needs) and their mapping to the BPMM sample. We hereby acknowledge that the weights refer to ideal values, obtained from the Delphi study, and might differ from weights that would be assigned by other organisations.

The output of the BPMM Smart-Selector concerns concrete BPMMs of our sample, that users can start using. However, the introduction chapter already suggested that BPMMs are likely to differ in quality. Hence, in order to guarantee the quality of the tool's output, we decided to screen the BPMMs before adding them to our BPMM Smart-Selector.

Particularly, two evaluation scores were calculated per BPMM, i.e. a selection score and a transparency score. The selection score copes with the 14 decision criteria that are included in the questionnaire, and assigns the obtained weights to BPMMs. On the other hand, a transparency score verifies whether all criteria considered in the Delphi study (i.e. without consensus to be excluded from the tool) are present in the BPMM design documents in order to be directly usable after selection. Based on these evaluation scores, nine BPMMs were omitted from the BPMM Smart-Selector. We emphasise that a similar quality check on the BPMM sample was not required for study 2, because the process capability areas were present in all BPMM design documents collected. Instead, for study 2, the quality check is rather implied into the BPMM classification, i.e. with cluster 1 being less comprehensive than clusters 2 or 3.

Finally, study 3 advised to strategically think through 14 decision criteria, listed according to their importance for our Delphi experts. It concerns those criteria that reached consensus to be included in the questionnaire of our BPMM Smart-Selector, and were used to calculate selection and transparency scores.

(1) Presence of capabilities: the business process capability areas to be assessed and improved.
(2) Architecture type: the possibility to define a road map per capability and/or a road map for overall maturity.
(3) Architecture details: the degree of guidance that a maturity model gives on your journey towards higher maturity.
(4) Type of business processes: whether the maturity model addresses specific process types (e.g. supply chains or collaboration processes) or can be applied to any process type.
(5) Rating scale: the type of data that is collected during an assessment.
(6) Data collection technique: the way information is collected during an assessment.
(7) Purpose: the purpose for which the maturity model is intended to be used.
(8) Validation: evidence that the maturity model is able to assess maturity and helps to enhance the efficiency and effectiveness of business processes.
(9) Number of business processes: the number of business processes to be assessed and improved.
(10) Assessment duration: the maximal duration of a particular assessment.
(11) Assessment availability: whether the assessment items and level calculation are publicly available (instead of only known to the assessors).
(12) Functional role of respondents: the explicit recognition to include people from outside the assessed organisation(s) as respondents.

(13) Number of assessment items: the maximal number of questions to be answered during an assessment.
(14) Direct costs: the direct costs to access and use a maturity model.

Subsequently, we list the less decisive criteria for BPMM selection, according to our Delphi experts. It concerns those criteria that did not reach consensus to be included in nor excluded from the BPMM Smart-Selector, and thus remain important to some extent. Such criteria were also used to calculate the transparency scores and can be optionally consulted in the BPMM Smart-Selector as additional information, but are not included in the questionnaire.

(15) Calculation of (maturity or capability) levels: the way the resulting (maturity or capability) levels are calculated.
(16) Representation of (maturity or capability) levels: the way the resulting (maturity or capability) levels are displayed.
(17) Number of assessed organisations: the number of organisations (i.e. autonomous legal entities) that are included in the assessment.
(18) Lead assessor: whether the assessment is led by an external, (quasi-) independent person, i.e. third party.
(19) Number of assessors: the number of assessors who are required to conduct an assessment.
(20) Business versus IT respondents: the explicit recognition to include IT people as respondents in the assessment.
(21) Number of (maturity or capability) levels: the number of (maturity or capability) levels that are defined.
(22) Labelling of levels: the way (maturity or capability) levels are labelled, i.e. what they indicate or explicitly refer to.
(23) External levels: the extent to which (maturity or capability) levels take into account possible relationships between individual organisations.
(24) Methodology: the way the maturity model was created.

We recall that the online version of our BPMM decision tool is available for free at: http://smart-selector.amyvanlooy.eu/.

3.2 Scientific Contributions

After having discussed our main findings, we now recapitulate the corresponding scientific contributions.

First, the common understanding of study 1 not only served this book, but can also help avoid confusion and inappropriate assumptions of BPMM researchers. For instance, it allows clear communication among scholars on the difference between business process maturity and capability. Also the BPMM definition can be used to evaluate whether a particular model is to be interpreted as a BPMM.

The findings are considered paramount given the many inconsistencies that we encountered in the BPMM sample, and the limited academic literature on BPMMs that still remained elusive so far. Finally, an important contribution outside the BPMM context is the proposed difference between BPM and BPO. The introduction chapter already mentioned that the business process literature mainly refers to BPM. BPO is hardly used, and if present, it is seen as a synonym for BPM. On the contrary, study 1 showed a funnel structure with BPM being part of BPO. We argued that BPM refers to characteristics of business processes, i.e. related to the traditional business process lifecycle, whereas BPO adds organisation-specific characteristics to BPM, i.e. to make the organisational culture and structure more process-oriented. This explicit distinction between BPM and BPO was required to emphasise a difference in BPMM scope. Moreover, it contributes to the business process literature in general by refining and linking two umbrella terms.

Study 2 contributes to BPMM research by presenting a conceptual framework of business process capability areas (i.e. with three clusters, six main areas, and 17 sub areas) and by identifying nine different types of business process maturity. At present, no consensus exists on the formal capability areas for mature business processes. Nevertheless, we hereby presented a solution that is supported by a large sample of 69 BPMMs. Another novelty is that we strongly relied on the underlying theories of the traditional business process lifecycle and organisation management theories, instead of only empirically examining capability areas as critical success factors for business (process) excellence. Furthermore, our conceptual framework can be used to evaluate the scope of existing BPMMs, i.e. depending on the covered capability areas, and to improve them, i.e. by developing the uncovered areas. Also the design of new BPMMs can be redirected accordingly. Moreover, we refined the earlier business process maturity types of de Bruin and Rosemann (2007). Our different maturity types were ranked according to their theoretical completeness. We also refined this ranking by suggesting that some organisations do not require the most complete BPMMs, e.g. depending on the available top management support, organisation size and the initial knowledge on process excellence. Nevertheless, study 2 explained that the maturity types allow a critical view on BPMMs in order not to be misled by the real BPMM names, i.e. based on the capability coverage and the number of processes addressed.

Finally, study 3 made scientific contributions regarding the design artefacts, the design foundation and the design methodology of the online decision tool, i.e. BPMM Smart-Selector. First, the design artefacts were based on the artefact types of March and Smith (1995):

- the conceptual model of the BPMM decision tool (i.e. constructs);
- the overview of decision criteria and their weights (i.e. model);
- the questionnaire that operationalises the decision criteria (i.e. model);
- the decision table that applies the questionnaire and selects the best matching BPMM (i.e. method);
- the decision tool, called BPMM Smart-Selector, that implements the decision table (i.e. the resulting instantiation).

Regarding the design foundation, study 3 extends the BPMM state-of-the-art research by gaining insight into the decision criteria, their trade-offs and weights. It also provided a formal way of calculating evaluation scores per BPMM, allowing a more critical view on the quality of BPMMs than study 2. Furthermore, the usefulness of our statistical clustering of study 2 was corroborated by a discussion among the Delphi experts. Particularly, when the experts were asked which main capability areas they want to assess and improve for their specific context, they grouped the areas similar to our statistical clusters. Regarding the design methodology, study 3 intended to build a theory (i.e. a design process or methodology) for building and testing BPMM decision tools by addressing the design theory components of Walls et al. (2004), and thus contributes to the (BPMM) design science literature. Finally, it was shown that the BPMM Smart-Selector can also be used by scholars, who want to apply a BPMM for their research.

3.3 Practical/Managerial Implications

Besides the scientific contributions, this book has several implications for organisations and their managers.

Study 1 mainly served academic purposes. However, its common understanding also supports practitioners in interpreting BPMMs and appropriately discussing BPMM results with peers.

More practical implications were derived from study 2. As business process capability areas are core to BPMMs, practitioners must be aware of the essential differences in capability coverage. Not surprisingly, the presence of capability areas was also evaluated as being the most important decision criterion for BPMM selection by the Delphi experts in study 3. A critical view on the capability coverage of BPMMs is not only a scientific contribution, but also a strong indication for practitioners on the direction of BPMMs, i.e. which capabilities need to be assessed and improved to increase business process maturity. Furthermore, advice was given on which statistical cluster can be chosen when.

Most practical implications are related to study 3, as our decision tool is primarily of practical use to managers wishing to select a BPMM. This study aggregated our research findings and made them freely available to organisations by means of a BPMM decision tool. Particularly by using our BPMM Smart-Selector, practitioners can find the BPMM that best fits their organisational needs, based on rational considerations. The BPMMs included in the BPMM Smart-Selector all passed a quality check, in order to enable a direct use after selection. Finally, the practical relevance of our BPMM Smart-Selector was demonstrated by conducting case studies. More specifically, the tool was first tested by four BPM students working for organisations of different sizes, i.e. as potential BPMM users. This pre-test resulted in some significant improvements to the way of working of our BPMM Smart-Selector. One important improvement was the possibility to skip questions in the questionnaire and thus to deviate from the sequence or weights proposed by the Delphi experts. Although the selected set of decision criteria was approved, i.e. no missing criteria identified, this finding also showed that the relevance among those criteria remains organisation-dependent. Afterwards, the tool was used in three real-life scenarios for which BPMM interest was formulated beforehand: (1) a business scenario at a profit organisation that did not yet use a BPMM, (2) a business scenario at a non-profit organisation that already used a BPMM, but that needed a BPMM with a better fit for purpose, and (3) an academic scenario with a scholar who wanted to apply a BPMM for research purposes. Data saturation was reached, as after pre-testing, the scenarios did not result in significant improvements to the tool. The empirical data collected in the scenarios also supported the hypotheses regarding our design process (i.e. the methodology used) and our design product (i.e. the BPMM Smart-Selector). Moreover, it's worth noticing that the term 'business' in BPMM can be interpreted broadly, i.e. BPMMs can be used by business people of both profit and non-profit organisations as well as by scholars studying those different organisations.

3.4 Research Limitations

While conducting our research, we mainly faced the following research limitations.

First, our findings strongly relied on a content analysis of the BPMM sample. Particularly, the BPMM classification (study 2) was based on a mapping to the business process capability areas, whereas the evaluation scores and the decision table of study 3 reflected a mapping to other BPMM characteristics. The introduction chapter explained the different measures that we took to guarantee objectivity or inter-subjectivity. Although the sample was repeatedly analysed over time, we did not statistically calculate inter- and intrarater agreements. For instance, we did not use a Cohen's kappa like we did when comparing the classification techniques in study 2 or when determining stability of the Delphi results in study 3. Nevertheless, our purpose was rather to draw general conclusions instead of focusing on the details of particular BPMMs. For instance, although analysed and written down as a potential paper, we deliberately not included our study on the descriptive statistics of the BPMM sample. Furthermore, study 2 used diverse classification techniques to validate the results and to reduce chance or randomness. Study 3 quantified the quality of individual BPMMs, but this quality check was proposed to improve the BPMM Smart-Selector. Moreover, the focus of study 3 was ultimately on the creation of a methodologically sound decision tool and on the strategic considerations during BPMM selection. The BPMM details in our database can still be easily updated or extended to other BPMMs, if necessary.

A second limitation concerns the scope of our BPMM sample. We collected BPMMs for supply chains or collaboration processes, besides generic processes, in order to acknowledge cross-organisational value chains. This decision was made to facilitate generalisation of our findings, and we uniformly analysed the whole sample. Nevertheless, we did not fully acknowledge the characteristics of specific process types. For instance, we did not elaborate on how supply chains differ from general business processes, or how cross-departmental BPMMs differ from cross-organisational BPMMs. Moreover, our findings were not validated by applying them to BPMMs for other specific process types, not included in our BPMM sample, such as software development processes. Instead, validity of the BPMM classification was primarily assured by (1) choosing a meaningful and statistically correct cluster solution after considering different algorithmic clustering methods, (2) guaranteeing stable results on both the complete and split dataset, and (3) combining exploratory and confirmatory classification techniques, i.e. cluster analysis and discriminant analysis. Hence, we rather focused on internal validity (i.e. to obtain accurate or credible results) and reliability (i.e. to obtain consistent or replicable results when repeatedly measured) than external validity (i.e. to transfer or generalise our findings to other specific process types, not included in the BPMM sample). Regarding the BPMM decision tool, internal validity follows from the Delphi study and case studies, and reliability is to some extent covered because no missing criteria were identified during the case studies. Other potential criteria, like those proposed in the design science literature, may be considered as not important or decisive enough for BPMM selection, given the sound methodological approach that was used. Also for this selection study, we argued that external validity is not demonstrated, but that our database structure should allow increasing the number of BPMMs and decision criteria to choose from without re-implementation. Similarly, the design process should be replicable to take new decision criteria into account.

Thirdly, this book contributes to the BPMM literature by gaining insight into the concept of business process maturity. Although the introduction chapter emphasised the importance of a maturity theory to overcome the criticisms, we do not pretend offering one. Instead, study 2 gave evidence for a conceptual framework with different capability areas and maturity types, as a first essential step towards a theoretical framework. The BPMM literature still lacked the necessary foundation to start from, and we have tried to overcome this initial barrier. We did not fully elaborate on the combinations of capability areas (i.e. maturity types) that contribute more to performance than others, or on the effect of the number of business processes addressed. Neither did we fully investigate the contextual factors affecting the cluster choice of organisations. Our study, however, suggested some potential contextual factors for future research, such as top management support, organisation size and the initial knowledge on process excellence.

Another main research limitation refers to the user's perspective that we took in study 3. We wanted to create a BPMM decision tool, instead of merely investigating BPMMs from design principles. However, the tool itself was created for users, but not by users. We derived initial decision criteria from the collected BPMM design documents and discussed final decision criteria in a Delphi study with BPM academics and practitioners, i.e. including potential users. We decided to start the Delphi study with initial criteria to obtain common ground among the Delphi participants and to acknowledge previous research results. We recognise that this approach could create a research bias. Nevertheless, the Delphi participants could give open comments at any time and they highly rated their overall satisfaction with the final decision criteria and the corresponding weights. Moreover, the importance of our BPMM Smart-Selector was supported by BPMM users in real-life case studies. Consequently, our design hypotheses were supported by the majority of both Delphi participants and BPMM users with satisfaction rates of 5–7 on a 7-point Likert scale. Such empirical evidence hardly suggests a bias, but of course, we cannot predict which decision criteria would have been resulted from a Delphi study without initial criteria.

3.5 Future Research

This section elaborates on the avenues that are worthwhile to consider in future research: (1) for building and testing a theory of business process maturity, (2) for enhancing existing BPMMs, and (3) for designing cross-organisational BPMMs. For each of the avenues, we indicate how they address the previously discussed research limitations.

3.5.1 Building and Testing a Theory of Business Process Maturity

This book advised on how an appropriate BPMM can be chosen by consulting international subject-matter experts. A logical next step would be to make this advice also dependent on a scientific theory. It concerns a maturity theory that predicts to which degree an increase in business process capabilities leads to higher maturity and higher business (process) performance, and which contextual factors must be taken into account.

Particularly, as a continuation of this book, future research can start from the conceptual framework of study 2 to build and test a theory of business process maturity. The proposed associations to be investigated in future research are summarised in Fig. 3.1. It would extend Fig. 1.6 from the introduction chapter with the identified business process capability areas and maturity types. Additionally, the effects between maturity types are to be examined, i.e. between the clusters and between the number of business processes addressed. For instance, how does BPO maturity affect intermediate BPO maturity and BPM maturity, and how does intermediate BPO maturity affect BPM maturity, or vice versa? Or does higher maturity of all business processes in the organisation also affect the maturity of individual business processes, or vice versa? Given that our identified maturity types also depend on the number of business processes addressed, a distinction is proposed between process performance, BPM performance

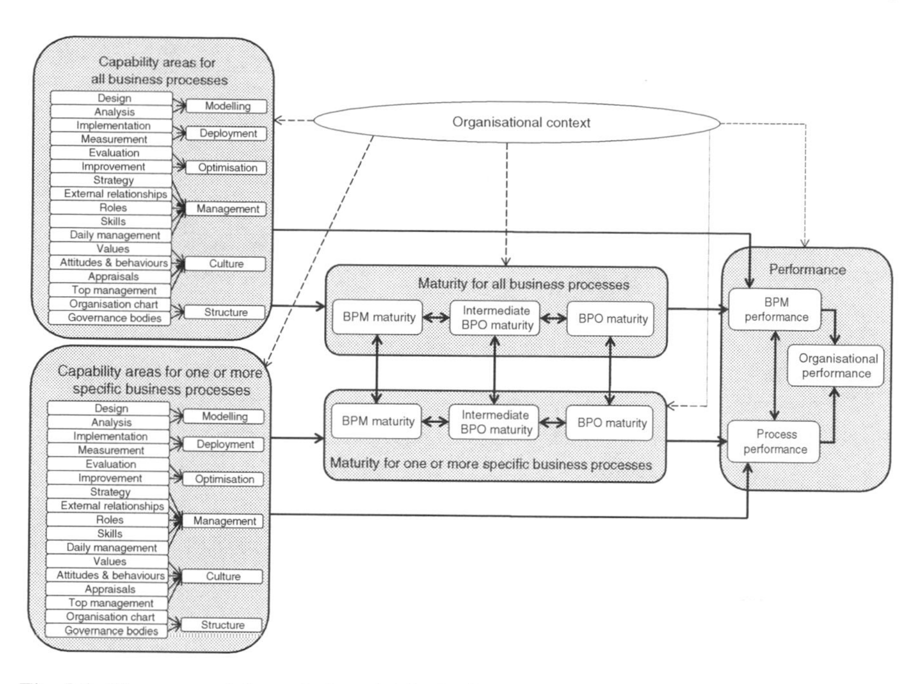

Fig. 3.1 The proposed theoretical model for business process maturity (future research)

(or process portfolio performance), and organisational performance. Particularly, the direct effects of improving one or more specific business processes on process performance must be investigated, as well as the direct effect of improving all business processes on BPM performance. Afterwards, the focus must switch to the indirect effects on organisational performance, and on the relationship between process performance and BPM performance. It can be demonstrated which combinations of capability areas (i.e. maturity types) contribute more to performance than others. For instance, do the additional efforts of BPO maturity significantly increase business (process) performance, compared to the basic efforts of BPM maturity? Or does it depend on contextual factors, such as the organisation size or market competitiveness? Study 2 and study 3 already raised the need to include contextual factors. For instance, which factors impact the cluster choice of organisations? The BPMM Smart-Selector can help collect possible contextual factors by requesting general data before navigating to the questionnaire. Hence, opportunities exist to further enhance the foundation of this proposed theoretical model, and test it accordingly.

More specifically, case studies can be conducted, followed by a large-scale survey (to be analysed using structural equation modelling). First, an additional literature study seems appropriate to elaborate on the different performance types and different contextual factors. The latter is based on contingency theories (Sect. 1.3, issue 3) and inspired by Delphi discussions (study 3, Delphi rounds 3 and 4). For instance, more advanced business process modelling might generally contribute to higher performance, but when controlled for organisation size, do the modelling efforts of smaller organisations also pay off? Or maybe smaller organisations are just not investing in business process modelling due to a cost-benefit analysis? Another contextual example is given for the type of business processes: does advanced business process modelling still pays off for support and management processes, i.e. which are not critical to customers? Hence, a business process theory must advice on optimal maturity levels, instead of blindly striving for the highest maturity levels, i.e. without under- or overinvesting in time, money, efforts, or resources. Furthermore, an additional literature study could predict which capability areas have a higher impact on performance than others, instead of uniformly assuming they all make similar performance contributions.

We hereby present some first ideas to operationalise business process maturity, its capability areas and performance contributions.

- For maturity, our BPMM Smart-Selector could be used to select BPMMs that are publicly available and free of charges. By using those BPMMs of our sample, all BPMM assessment items must also be included into the intended survey, risking a long survey. Another option is to collaborate with owners of existing BPMMs, i.e. they could provide a list of organisations and their current maturity levels and future research then sends a survey (i.e. without questioning maturity) to those organisations. In order to avoid bias of particular BPMMs and for reasons of generalisation, it is paramount that each maturity type is measured by multiple BPMMs. Furthermore, the relationships between the maturity types can only be investigated if their difference in scope is unambiguously explained in the intended survey. This means that if the same business people would use BPMMs for measuring different maturity types, they also consider different subsets of business processes. Pre-tests must assure a clear terminology, and check how business people experience the distinction between one, more, and all processes (i.e. as proposed by the assessment items of existing BPMMs). Particularly regarding BPMMs for one or more specific processes, process boundaries might be difficult to define and sub processes may not be experienced as such. If necessary, future research could opt to combine BPMMs addressing one and more specific business processes, in order to focus on their unambiguous difference with BPMMs addressing all business processes in the organisation (Fig. 3.1). Finally, as BPMMs vary on the number of maturity levels, an organisation's final level per maturity type can be obtained by standardising its maturity level per BPMM (i.e. dividing the current level by the maximal level), and then averaging those standardised values among BPMMs.
- Capability areas must be independently measured from maturity, because using similar assessment items for both concepts would not allow measuring relationships between them. This contradicts to the way BPMMs works, but allows verifying the implicit assumptions typically made by BPMMs, and further validating the found clusters or maturity types. Therefore, maturity can be measured by existing BPMMs (i.e. each covering a certain number of sub areas with model-specific assessment items), whereas the theoretically identified sub

areas can be measured based on the literature (study 2). However, one must also be aware that creating a scale with new assessment items risks creating yet another BPMM. Particularly, the gradations on a 5-point or 7-point Likert scale are closely related to maturity levels and capability levels. Alternatively, the sub areas could be measured with binary data, and then summed up to reflect the main areas. With binary data, we mean a checklist that concretises the presence of each sub area, according to the literature study. Pre-tests must assure that the capability areas are properly captured without losing much detail. Furthermore, factory analysis could be used to investigate whether more dimensions are present than our six main capability areas, e.g. must the 'culture' main area with four sub areas be reorganised into two separate areas (let's say 'values, attitudes and behaviours' and 'rewards and top management')? If the pre-tests, however, show that a checklist does not properly collect the necessary data, future research could opt for a minimal 3-point scale (i.e. disagree, partly agree, agree), a nominal measurement level or an open-ended question to better reflect all or some sub areas.

- Finally, much literature exists on organisational performance and process performance. Validated scales are preferred for organisational performance, such as the assessment items of business excellence models (like EFQM) or other academic studies (McCormack 2007; Skrinjar et al. 2008). Examples of process metrics are given in (Gonzalez et al. 2010). Future research must investigate whether appropriate scales also exist for BPM performance (or process portfolio performance), and if not, create one. For reasons of comprehensiveness, objective data (i.e. quantitative metrics) can be combined with subjective data (e.g. perceptions on performance gains due to process investments). For instance, quantitative metrics ensure some degree of objectivity in the data set, but perceptions can prove more easily whether an increase in organisational performance is also ascribed to process improvements.

With this research avenue, we try to respond to the first and third research limitations. Particularly, the proposed activities would not rely on the content analysis of our BPMM sample, but on case studies and surveys. Finally, the third research limitation will inherently be addressed while building and testing a theory of business process maturity.

3.5.2 Enhancing Existing BPMMs

Besides theory development, the findings within this book can be directly used to enhance existing BPMMs, for example:

- to make their capability coverage more comprehensive by reflecting all capability sub areas;
- to make their design documents more transparent by reflecting all BPMM characteristics.

Enhancing existing BPMMs implies coping with the first research limitation regarding the content analysis of BPMM design documents. Analysing BPMMs through descriptive statistics allows going into the details of other BPMM characteristics than business process capability areas. Also the second research limitation can be addressed by comparing BPMMs for generic business processes to BPMMs for specific process types.

For instance, a potential research avenue is to elaborate on the (dis)advantages of BPMMs for generic processes, compared to BPMMs for specific process types. For instance, the differences and similarities between generic processes and supply chains could be studied. Our study suggests that all BPMMs have the same foundation, but with particular translations. For instance, during content analysis, we experienced that BPMMs for supply chains pay more attention to the main capability areas of 'optimisation' and 'management', and this with specific technologies like track-and-trace systems to visualise the material flow, vendor-managed inventory (VMI), and collaborative planning, forecasting and replenishment (CPFR). The findings could supplement the trade-offs that we mentioned in our BPMM Smart-Selector for choosing between a generic BPMM and a domain-specific BPMM.

An additional research issue concerns the labelling of maturity levels. The level labels in our BPMM sample referred to (1) business process optimisation (e.g. initial, managed, standardised, predictable, innovating) (OMG 2008), (2) business process management (BPM) (e.g. BPM initiation, BPM evolution, BPM mastery) (Rummler-Brache Group 2004), or (3) business process integration (e.g. ad hoc, defined, linked, integrated) (McCormack and Johnson 2001). These three types of labelling appear to describe the strategy or goal of a given BPMM, albeit by improving similar capabilities. In our sample, BPMMs about one business process have optimisation levels or management levels, BPMMs about more business processes have one of the three label types, and BPMMs about all business processes mostly have management levels or integration levels. Hence, the level label choice may be guided by the number of business processes addressed. Although all label types seem appropriate, no BPMM in our sample simultaneously combines all label types for the same set of processes. Stimulated by open comments during the Delphi study for the criterion 'labelling of levels', future research could investigate whether all labels are interchangeable and just window-dressing for the numerical labels (e.g. whether level 3 is equal for all BPMMs), or whether a single BPMM ideally combines all level label types, i.e. by refining into three maturity lifecycles as end goals of BPMMs (besides the capability levels per capability area). For instance, given a particular assessment of all capability areas, your organisation is considered mature (i.e. capable of performing excellently) for improving certain processes with level 4, managing those processes with level 3 and integrating them with level 2. We must, however, note that optimisation and management are also main capability areas with associated capability levels, and that external integration might not be a goal for all processes. What are appropriate labels to characterise each level? Does it matter?

Given that BPMMs mainly have five lifecycle levels, most organisations are currently situated on level 2 or 3 (e.g. in BIS, CSC, FAA, HAR2, MCL, ORA, PMG, ROH, ROS, RUM, SEI), with some above (e.g. in MCC2) or below (e.g. in ARY, GAR2, RIV, SAP). Some of those organisations are also expected to evolve towards higher maturity levels in the coming years or decades. Nevertheless, we recall that not all business processes must reach the highest maturity levels. Managers should decide themselves which processes are important enough to invest up to the highest maturity levels. The intended theory of business process maturity could facilitate their choice. Meanwhile, we also encourage research on BPMMs as a fit concept based on contingency theories (i.e. with specific tracks depending on contextual factors) instead of the current BPMMs that provide similar road maps for all organisations (i.e. one model fits all). The introduction chapter referred to such BPMMs as 'situational' models. In our sample, half of the BPMMs pretend to be applicable to all organisations, but only a few support this statement by evidence. BPMMs are commonly applied by medium and/or large organisations, which usually operate with more complex business processes than micro or small organisations. Studies have empirically examined that larger organisations benefit more from a process-oriented approach (El Emam and Birk 2000; Gustafsson et al. 2003; Rout et al. 2007). Above all, other empirical evidence exists for the varying effects of BPM techniques on organisational performance, which may also depend on the industry (Ittner and Larcker 1997). To our knowledge, BPMMs for small organisations are still lacking nowadays.

Finally, the question remains to which degree BPMMs must cope with IT in their road map to enable business processes. During content analysis, we experienced that the majority of BPMMs prescribe IT to reach higher maturity. This degree, however, varies from (1) general IT, e.g. merely mentioning the terms 'hardware' or 'software', to (2) specific technologies, e.g. process-aware information systems or enterprise resource planning (ERP), and (3) specific vendor tools, e.g. ARIS. For our BPMM sample, IT was mostly prescribed to improve the traditional business process lifecycle, i.e. the main capability areas of 'deployment' (82.6 %), 'optimisation' (75.4 %), and 'management' (60.9 %), but with a much smaller role for 'modelling' (40.6 %). A possible explanation why 'modelling' lags behind might be the typical IT gradation (from no design, manual design, design with MS Office to design with a process-aware information system), whereas other areas show more IT variations. Further examination could reveal whether IT modelling tools are likely too complex or expensive to be imposed by BPMMs. Again, it is questionable whether the collected BPMMs are applicable to all organisations with respect to the IT they recommend. For instance, large product organisations may benefit more from automation than small service organisations. Therefore, we assume that a generic BPMM must rather be IT-neutral or generally IT-focused, i.e. without prescribing specific technologies or vendor tools. Nevertheless, a study could analyse the appropriate role of IT, based on empirical research and multidisciplinary theories regarding social informatics, IT acceptance, business-IT alignment, etc. The findings could also contribute to the intended theory of business process maturity.

3.5.3 Designing Cross-Organisational BPMMs

A final avenue for future research considers our second research limitation by elaborating on the differences between cross-departmental and cross-organisational BPMMs.

Business processes are increasingly crossing the borders of organisations, for instance due to e-business, e-government or business process outsourcing. Nonetheless, they are typically studied within a single organisation. Besides this conventional BPM approach, recent research is examining processes that cross the organisational boundaries, called 'collaborative BPM' (Werth et al. 2009) or 'network BPM' (Grefen et al. 2009). During this research, the cross-organisational approach of BPM was still in its infancy and mainly limited to modelling issues. Hammer (2010) explains that '*some companies have been working on these processes, but we lack models for their governance and management. Who is the process owner? How should benefits be allocated? What are the right metrics?*' (p. 15). Descriptive statistics on the BPMM sample confirm that also BPMMs are generally not well prepared for the future of cross-organisational collaboration and interoperability (Table 3.1).

As explained in Table 3.1, BPMMs are mainly cross-departmental, possibly with external collaboration on the highest maturity levels. Most assessments are conducted by the managers and staff of one organisation. Nevertheless, information obtained from external parties may add interesting perspectives for (future) cross-organisational collaboration and policy acceptance. The limited number of cross-organisational BPMMs principally focuses on optimisation levels, but not on lifecycles regarding management or integration efforts. This finding reflects the practical difficulty of advanced collaboration among autonomous parties. However, it strongly indicates a gap in the literature (especially regarding supply chains and collaboration processes), as BPMMs inherently aim to facilitate complex improvements.

But how can BPMMs properly address cross-organisational collaboration? Especially regarding BPMMs for generic business processes, it may seem logic that organisations start with improving their internal working, before adding more complexity by extending towards partners. However, those maturity models assume that higher levels with external collaboration may not be reached before all business processes addressed are internally integrated. On the other hand, an integrated internal structure does not necessarily leads to external integration. Moreover, not all customers and suppliers are necessarily and permanently involved in an end-to-end business process (Lambert and Cooper 2000). Therefore, based on empirical evidence, CAM3 distinguishes three integration types as distinct maturity lifecycles, without calculating overall maturity: internal integration, forwards with customers, and backwards with suppliers. This idea is conform to the integration arcs of Frohlich and Westbrook (2001), although the latter do not provide lifecycle levels. Another solution to cope with cross-organisation collaboration is given by ISO, which includes all focal organisations involved to improve

Table 3.1 The BPMM characteristics related to cross-organisational business processes (N = 69)

BPMM characteristic	Descriptive statistics
Number of assessed organisations	88.4 % are limited to one focal organisation, of which most BPMMs for supply chains and half of the BPMMs for collaboration processes. For these maturity models, the perspective of only one partner or collaborating organisation is taken while assessing maturity in the supply chain or collaboration
Functional role of respondents	86.96 % limit respondents to managers and staff of a single organisation
External levels	Of the 61 BPMMs with one focal organisation:
	• 31.15 % have no notion of external organisations in their lifecycle levels
	E.g. initial, managed, standardised, predictable, innovating processes for one organisation (OMG)
	• 60.65 % have notion of external organisations as from the highest lifecycle levels, but start by improving their internal way of working
	E.g. ad hoc, defined, linked, integrated, extended processes (MCC2)
	• 8.20 % have notion of external organisations on all lifecycle levels, and are thus also considered cross-organisational
	E.g. performed, managed, standardised, and innovating processes for more organisations simultaneously, but assessed from the perspective of one organisation (ESI2)
	Finally, the 8 BPMMs allowing multiple focal organisations are inherently cross-organisational
Labelling of levels	No BPMM with multiple focal organisations focuses on process integration in their level labels
	7 of the 8 BPMM allowing multiple focal organisations have level labels that primarily focus on business process optimisation, and one BPMM has management level labels

business processes along their lifecycle, i.e. by focusing on process optimisation without discussing process integration or management challenges. Future research is required to examine how BPMMs can be broadened towards multiple actors, e.g. how to deal with shared process ownership (Larsen and Klischewski 2004)?

3.6 End Notes

Research on business process maturity has still a great potential to be explored. Therefore, we would like to encourage other researchers to join us facing the identified challenges. This work resulted in a better BPMM foundation based on a unique and large sample of existing BPMMs. We are now one step closer to a theory of business process maturity, but a long way is still ahead of us.

References

de Bruin T, Rosemann M (2007) Using the Delphi technique to identify BPM capability areas. Proceedings of the 18th Australasian conference on information systems, Toowoomba, pp 642–653, 5–7 Dec 2007

El Emam K, Birk A (2000) Validating the ISO/IEC 15504 measure of software requirements analysis process capability. IEEE Trans Softw Eng 26(6):541–566

Frohlich MT, Westbrook R (2001) Arcs of integration: an international study of supply chain strategies. J Oper Manag 19:185–200

Gonzalez LS, Rubio FG, Gonzalez FR, Velthuis MP (2010) Measurement in business processes: a systematic review. Bus Proc Manag J 16(1):114–134

Grefen P, Mehandjiev N, Kouvas G, Weichhart G, Eshuis R (2009) Dynamic business network process management in instant virtual enterprises. Comput Ind 60(2):86–103

Gustafsson A, Nilsson L, Johnson MD (2003) The role of quality practices in service organizations. Int J Serv Ind Manage 14(2):232–244

Hammer M (2010) What is business process management? In: vom Brocke J, Rosemann M (eds) Handbook on business process management, vol 1. Springer, Berlin Heidelberg, pp 3–16

Ittner CD, Larcker DF (1997) The performance effects of process management techniques. Manage Sci 43(4):522–534

Lambert DM, Cooper MC (2000) Issues in supply chain management. Ind Mark Manage 29:65–83

Larsen MH, Klischewski R (2004) Process ownership challenges in IT-enabled transformation of interorganizational business processes. Proceedings of the 37th Hawaii international conference on system sciences, Hawaii, pp 1–11

March ST, Smith GF (1995) Design and natural science research on information technology. Decis Support Syst 15(4):251–266

McCormack K (2007) Introduction to the theory of business process orientation. In: McCormack K (ed) Business process maturity, theory and application. Booksurge Publishing, South Carolina, pp 1–18

McCormack K, Johnson WC (2001) Business process orientation: gaining the e-business competitive advantage. St. Lucie Press, Florida

OMG (2008) Business process maturity model (BPMM). Version 1.0. http://www.omg.org/spec/BPMM/1.0/PDF. Accessed 2 Dec 2009

Rout TP, El Emam K, Fusani M, Goldenson D, Jung H-W (2007) SPICE in retrospect: developing a standard for process assessment. J Syst Softw 80:1483–1493

Rummler-Brache Group (2004) Business process management in U.S. firms today. http://rummler-brache.com/upload/files/PPI_Research_Results.pdf. Accessed 23 June 2010

Skrinjar R, Bosilj-Vuksic V, Stemberger MI (2008) The impact of business process orientation on financial and non-financial performance. Bus Proc Manage J 14(5):738–754

Walls JG, Widmeyer GR, El Sawy OA (2004) Assessing information system design theory in perspective: how useful was our 1992 initial rendition? JITTA: J Inf Technol Theory Appl 6(2):43–58

Werth D, Walter P, Loos P (2009) Distribution and composition of collaborative business processes through peer-to-peer networks. Proceedings of the BPM 2008 workshops. Springer, Milano, pp 597–608

About the Author

Dr. Amy Van Looy

Dr. Amy Van Looy holds a Ph.D. in applied economics. She is assistant professor at the Faculty of Economics and Business Administration of Ghent University (Belgium). Before entering academia, Amy worked as an IT consultant (i.e., mainly business and functional analyst) for large e-government projects. Her research focuses on business process maturity and capabilities in public and private organisations, by considering the traditional process lifecycle as well as the organisational culture and structure. Other research interests include business process integration and business process modelling. Her research and publications can be accessed at http://www.amyvanlooy.eu/. The best way to contact Amy is via mail (info@amyvanlooy.eu). You can subscribe to her tweets at http://twitter.com/AmyVanLooy.

A. Van Looy, *Business Process Maturity*, SpringerBriefs in Business Process Management, DOI: 10.1007/978-3-319-04202-2, © The Author(s) 2014